Biological Applications of Anti-Idiotypes

Volume II

Editor

Constantin A. Bona, M.D., Ph.D.
Professor of Microbiology
Mount Sinai School of Medicine
New York, New York

CRC Press
Taylor & Francis Group
Boca Raton London New York

CRC Press is an imprint of the
Taylor & Francis Group, an **informa** business

First published 1988 by CRC Press
Taylor & Francis Group
6000 Broken Sound Parkway NW, Suite 300
Boca Raton, FL 33487-2742

Reissued 2018 by CRC Press

Library of Congress Cataloging-in-Publication Data

Biological applications of anti-idiotypes.

   Bibliography: p.
   Includes indexes.
   1. Immunoglobulin idiotypes. I. Bona, Constantin A.
QR186.7.B565   1988   616.07'93   87-20833
ISBN 0-8493-6941-X (v. 1)
ISBN 0-8493-6942-8 (v. 2)
ISBN 0-8493-6940- 1 (set)

A Library of Congress record exists under LC control number: 87020833

ISBN 13: 978-1-315-89112-5 (hbk)
ISBN 13: 978-1-351-07022-5 (ebk)

Visit the Taylor & Francis Web site at http://www.taylorandfrancis.com and the
CRC Press Web site at http://www.crcpress.com

PREFACE

The most demanding aspect of editing a book dealing with a specialized scientific topic is covering a vast body of literature through which as much available information as possible on the subject is obtained. However, it is also undeniable that much is gained from this mandatory exercise, not only to have a deep insight into the intimate dialectic of a science and the evolution of its concepts, but also the position which a particular discipline occupies among the realm of the general sciences of nature is revealed.

Comparatively speaking, immunology is a young science. In contrast to many sciences conceived 25 centuries ago under a Hellenistic sky dominated by the postulate that the universe is governed by rules accessible to the human mind, immunology was born at the end of the last century. In spite of its youth, the rapidity at which progress in the understanding of a system was attained in immunology is rivaled in very few fields. Such an accomplishment is attributable to the application of the same principles of creativity which permitted comparable achievements in other disciplines.

The strategy of immunological research was based on analysis: namely, the dissection of the components of the immune system and synthesis of information into the formulation of concepts. Often, investigators possessed sufficient insight and understanding of their systems to allow them to cumulate concepts even before their analytical studies were completed.

Other elements have also contributed to the rapid progress in immunology. First, research tactics were continuously adapted to take advantage of recent technological advances. There is no doubt that the success of experimental work is directly related to the technological approaches employed. Second was the collective character of the research. This is probably related to the interdisciplinary character of the science. It is compulsory for an immunologist to not only keep abreast of the technology but also to be familiar with the principles of biology and medicine to understand the organization and function of the immune system and the implication of any alteration of these functions in various diseases. Of course, it is not possible to minimize the role which intuition plays in contributing to brilliant discoveries.

Intuition is an individual event which allows one to find the most suitable approach leading to a major scientific breakthrough.

I do not believe that intuition is due to chance. Rather, it represents an excellent chronological juxtaposition between the accumulation of a critical amount of analytical findings and the erudite ability of conceptual synthesis culminating in a giant strike forward. Only well-prepared persons can have such illuminating intuition.

The text of a scientific book requires a few basic elements which include the presentation of findings, concepts, and theories. The presentation of findings is an essential component of this type of publication when they are the fruit of correctly executed experiments that can be assigned to the category of "eternal truths" which are no longer susceptible to transformation or improvement. However, it should be recognized that there is a legion of findings whose presentation is just a means to achieve a more distinct goal.

Indeed, one may ask whether a histologist who spent his life peering through a microscope at slides made from a horse, in fact, knows how the horse looks or runs.

To study the lymphocyte (its structure and physiology), biosynthetic products, and genes is not a goal in itself; such studies achieve a meaningful aim only by the collective integration of all the knowledge which they provide into a coherent assembly. Newton perhaps provided the best example of this in science by understanding that the physical laws governing the fall of a stone also govern the movement of stars.

The presentation of concepts requires a keen selection process in order to retrieve from the reams of available information these seminal findings supporting the concepts as well as the exceptions. Two features should not be forgotten in this regard. First, concepts are not everlasting since they are formulated from a limited amount of available information.

It is intriguing to follow the rapid transformation of concepts and models in immunology as new findings are brought to light. The second is to place in the proper perspective the anthropocentric input in concepts which are aimed at defining objective laws to rationalize the phenomena. I feel that numerous concepts and models in immunology have too much anthropomorphic input; sometimes we must resist the temptation to formulate orderly rules for governing a system when our true mission is only to unravel them.

The third element is the presentation of theories, classifying the findings into categories in a coherent order, and integrating the concepts. Theories are also not timeless since the curiosity of the experimental investigator continuously generates new findings by challenging the established theories. Thus, the writing of a scientific book is just a beginning and each beginning is imperfect.

The goal of this book was to present findings which are supportive of the concept that the words of the idiotype dictionary and likewise self-recognition are essential for communication between lymphocyte clones while some unfaithful (''infidel'') copies of the words contained in the idiotype dictionary perhaps play a role in the functional regulation of nonlymphoid cells.

The notion of self-recognition as a regulatory force was very important for my understanding of idiotype regulation. Since immunology evolved as a discipline of microbiology, a major paradigm which dominated immunological theories was selection during phylogeny of lymphocytes to ensure the defense mechanisms. Therefore, the immune system is oriented solely towards the recognition of foreign antigens. We wanted to point out that idiotype mediated self-recognition is an important regulatory force within the immune system.

THE EDITOR

C. A. Bona, M.D., Ph.D., is Professor of Microbiology at Mount Sinai School of Medicine, New York.

Dr. Bona received his M.D. and Ph.D. degrees in Medical Science from the Faculty of Medicine, Bucharest, and his Docteur in Sciences Naturelles degree from Paris University.

Dr. Bona has published more than 150 papers in various national and international journals, 29 chapters in various monographs, 6 sections in textbooks, has edited 2 monographs, and has written 2 books on idiotypes.

His major research interest is in the idiotypy field in which he discovered the naturally occurring idiotype-specific T cells, regulatory idiotopes, epibodies, and the first demonstration that an anti-Id antibody can elicit (instead of an antigen) an immune response against a bacterial antigen. This observation stimulated the research in the utilization of idiotypes as vaccines.

CONTRIBUTORS

Volume II

E.J. Bunschoten, Dr.
Investigator
National Institute of Public
Health and Environmental Hygiene
Bilthoven
The Netherlands

Jeffrey Cossman, M.D.
Senior Investigator
Hematopathology Section
National Cancer Institute
National Institutes of Health
Bethesda, Maryland

Donard S. Dwyer, Ph.D.
Assistant Professor
Neuropsychiatry Research Program
University of Alabama at Birmingham
Birmingham, Alabama

Hildegund C.J. Ertl, M.D.
Associate Professor
The Wistar Institute
Philadelphia, Pennsylvania

Nadir R. Farid,
Professor of Medicine
Division of Endocrinology
Department of Medicine
Health Sciences Center
St. John's, Newfoundland
Canada

Robert W. Finberg, M.D.
Associate Professor of Medicine
Department of Infectious Diseases
Dana Farber Cancer Institute
Department of Medicine
Harvard Medical School
Boston, Massachusetts

Denis Glotz, M.D.
Institut National de la Santé et
de la Recherche Scientifique
Paris, France

Edmond A. Goidl, Ph.D.
Associate Professor
Department of Microbiology and
Immunology
University of Maryland School of
Medicine
Baltimore, Maryland

Dorothee Herlyn, D.V.M.
Associate Professor
Wistar Institute
Philadelphia, Pennsylvania

Donald West King, M.D.
Richard T. Crane Professor of Pathology
School of Medicine
University of Chicago
Chicago, Illinois

Hilary Koprowski, M.D.
Director Wistar Institute
Philadelphia, Pennsylvania

Ann Kari Lefvert, M.D., Ph.D.
Associate Professor
Department of Internal Medicine
Karolinska Institute
Stockholm
Sweden

A.D.M.E. Osterhaus, D.V.M., Ph.D.
Head, Department of Immunology
National Institute of Public Health and
Environmental Hygiene
Bilthoven, The Netherlands

Mark Raffeld, M.D.
Senior Staff Fellow
Hematology Section
Laboratory of Pathology
National Cancer Institute
National Institutes of Health
Bethesda, Maryland

Joy Rogers
Division of Immunology
Medical Biology Institute
La Jolla, California

Kathryn E. Stein, Ph.D.
Research Chemist
Division of Bacterial Products
Office of Biologics Research and Review
Food and Drug Administration
Bethesda, Maryland

Nicole Suciu-Foca, Ph.D.
Professor of Pathology
College of Physicians & Surgeons
Columbia University
New York, New York

F.G.C.M. UytdeHaag, D.V.M., Ph.D.
Senior Scientist
Department of Immunobiology
National Institute of Public Health and
 Environmental Hygiene
Bilthoven
The Netherlands

K. Weijer, D.V.M., Ph.D.
Senior Scientist
Netherland Cancer Institute
Amsterdam
The Netherlands

Linda Woo
Research Technician II
Department of Infectious Diseases
Dana-Farber Cancer Institute
Boston, Massachusetts

Maurizio Zanetti, M.D.
Associate Professor
Department of Medicine
University of California at San Diego
San Diego, California

TABLE OF CONTENTS

Volume I

TABLE OF CONTENTS

Volume II

Chapter 1

ANTI-IDIOTYPES AS BACTERIAL VACCINES

K. E. Stein

TABLE OF CONTENTS

I. INTRODUCTION

Both the humoral and cellular immune systems are involved in protecting the host against bacteria. In most cases, the presence of serum antibodies directed against appropriate bacterial antigens is sufficient to prevent disease.[1-3] For a minority of bacteria, although antibodies may play some role in host defense, T-cell immunity is the primary mechanism by which the host is protected.[4] These considerations are of paramount importance in the development of efficacious vaccines, whether they are of the conventional type, i.e., consist of a bacterial product(s) (I would include those products actually made by the organism itself as well as those produced in the laboratory by either synthetic or recombinant DNA technology in this category), or are of a type not directly involving a bacterial product, such as an anti-idiotype (anti-Id). I will briefly discuss some examples of bacterial antigens which stimulate protective immunity in man. Conventional bacterial vaccines were recently reviewed in a book edited by Germanier[5] which should be consulted by those interested in bacterial immunity.

Among the bacterial antigens that stimulate protective antibodies are both proteins and polysaccharides. Examples of protein antigens that are known virulence factors are bacterial toxins, such as tetanus[6] and diphtheria[7] toxins. Both tetanus and diphtheria toxoids have been demonstrated to be safe and to induce protective immunity. For other bacteria that secrete toxins, such as *Bordetella pertussis*[8] and *Vibrio cholerae*,[9] the use of the toxoid alone as a protective vaccine has not yet been established, although for *B. pertussis* a heterogenous acellular preparation has recently replaced the whole cell vaccine in Japan.[10]

A high proportion of invasive bacterial infections is caused by encapsulated organisms such as *Neisseria meningitidis*,[11] *Haemophilus influenzae*,[12] and *Streptococcus pneumoniae*.[13,14] Antibody to the polysaccharide confers protective immunity,[1-3] however, most polysaccharides are poorly immunogenic in infants under the age of 2 years.[12,15-18] Thus, those at highest risk of disease cannot be protected by the available polysaccharide vaccines. Enhanced immunogenicity has been achieved, however, by coupling either polysaccharides[19-23] or oligosaccharides[24-27] to protein carriers. Significant antihemophilus type b (HIB) capsule antibody responses have been observed in infants over the age of 6 months immunized with thymus-dependent conjugates.[23,28-31]

T-cell immunity is required for protection of the host from bacteria that multiply within cells, such as *Mycobacterium tuberculosis*,[32] *M. leprae*,[33] *Listeria monocytogenes*,[34] and *Salmonella typhii*.[35,36] Experimental models of such infections have demonstrated that immunity can be passively transferred by T cells and not by serum antibody, although in the infected host, antibody may play a modulating role.[37] It would seem that organisms for which antibody confers protection might be the most amenable to the approach of an anti-Id vaccine, but those bacteria requiring T-cell immunity can also be considered, as will be discussed below.

Conventional vaccines have proven efficacious since the time of Jenner. Why, then, should we develop anti-Id for general-use vaccines? This question is particularly relevant today when it should be possible to clone the gene or sequence and synthesize, *de novo*, almost any bacterial product. In the subsequent sections, I will discuss why anti-Id as vaccines may be useful and review the literature on anti-Id as they relate to bacterial antigens. The term anti-Id will be used to refer to both polyclonal and monoclonal antibodies and I will only distinguish between them where necessary. Similarly, I will not distinguish between anti-Id of the "classic" type ($Ab_{2\alpha}$) and anti-Id of the "internal image" type ($Ab_{2\beta}$)[38] except where necessary. Other chapters in this volume discuss these distinctions in depth.

II. WHY USE ANTI-ID?

A. Introduction

Anti-Id vaccines represent a new approach to the stimulation of protective immunity. The

Table 1
WHY USE ANTI-ID?

They are normal host components
They are effective in situations where antigen is not
    Neonates
    Activation of silent clones
    Immunodeficiencies
They can be used when antigen is not available or is harmful

concepts involved in the design of such vaccines are based on the network theory of Jerne[39] and were first elaborated by Nisonoff and Lamoyi[40] with respect to their application to vaccines. Although anti-Id are not expected to replace conventional vaccines in most instances, this approach can lead to safe and efficacious vaccines in cases where conventional vaccines have not yet succeeded. The ability to purify and completely characterize an anti-Id, particularly a monoclonal antibody, gives this type of vaccine an advantage over some bacterial vaccines that consist of heterogeneous material or even whole cells. Table 1 summarizes the indications for the use of anti-Id vaccines.

B. Anti-Id Are Normal Host Components

Anti-Id vaccines, in practice, may not be very different from antigen vaccines except that they enter the regulatory pathway of the immune system at a different place. During the course of an immune response, following a single immunization with a bacterial PS, oscillating waves of antibody (idiotype [Id]) and anti-Id have been observed. These auto-anti-idiotype (auto-anti-Id) responses have been reported to bacterial levan (BL)[41] and to dextran[42,43] in inbred mice. They have also been found following hyperimmunization with *Micrococcus lysodeikticus* in outbred rabbits.[44] Similar anti-Id responses to the hapten, TNP, coupled to a semisynthetic PS have also been reported.[45] Because the immune system is constantly exposed to environmental PS, many of which bear cross-reactive determinants, the observed oscillations of Id and anti-Id likely represent the normal regulatory mechanism by which the overproduction of specific antibodies is prevented. Anti-Id antibodies have also been observed following immunization with a protein antigen. Geha reported both serum auto-anti-Id antibodies[46] and peripheral blood B cells[47] specific for antitetanus antibodies following a routine booster immunization with tetanus toxoid. Similar findings were also reported by Saxon and Barnett.[48]

It has not yet been established if the natural production of auto-anti-Id antibodies is related to the degree of clonal restriction of the antibody response, however, for PS antigens this may be an important factor. These antibody responses are generally pauciclonal, as has been well documented for murine[49-52] and rabbit[53] antibody responses. Relatively little is known about the human antibody repertoire for bacterial antigens, although it has been more than 20 years since Kunkel et al.[54] reported the existence of individual antigenic determinants (Id) on human anti-PS antibodies specific for BL and dextran. The studies from Kunkel's laboratory were the first reports that anti-PS antibody responses were both Id[54] and isotype[55] restricted. Recently, Insel et al.[56] reported that the IgG response to HIB PS in man is restricted, as observed by isoelectric focusing, and that there is apparent sharing of clones between unrelated individuals. A V_H-associated Id has been described on human antitetanus antibodies.[57] The inheritance of this Id was demonstrated in family members and it was shown that the rabbit anti-Id cross-reacted with the antitetanus antibodies of one of four unrelated individuals tested. More recently, a cross-reactive anti-Id was described that showed some reactivity with the antitetanus antibodies of 15 of 19 unrelated individuals.[58]

It should be possible with improved methodology for the production of human hybridomas to rescue and immortalize in vitro human monoclonal anti-Id produced during the course of immunization or following an infection. The use of human antibodies may not be necessary, however, if chimeric proteins consisting of mouse variable regions and human constant regions[59] prove to be safe in humans.

C. Anti-Id Are Effective in Situations Where Antigen Is Not

Although the capacity to respond to protein antigens is present at birth, the capacity to respond to PS antigens develops late in ontogeny. In mice this begins at 3 to 4 weeks of age and reaches adult levels at 10 to 12 weeks of age.[41,60-63] In man, responses to PS antigens have been observed at 3 to 6 months of age, reaching adult levels at 2 to 5 years.[12,16,64] Because the response to PS antigens is thymus independent,[61] efforts at stimulating anti-PS immunity in neonates have been concerned with the conversion of these responses to thymus-dependent forms. As discussed above, one way to accomplish this is to covalently couple poly- or oligosaccharides to protein carriers. Another way is to use an anti-Id vaccine. The ability of anti-Id to induce Id has been shown in some systems to require the presence of the thymus.[65] As will be discussed below, anti-Id does prime neonatal mice for a subsequent response to PS, whereas PS given at birth fails to prime for an anti-PS response. It is generally accepted that protein antigens stimulate a memory response, whereas PS antigens do not.[66,67] Thus, anti-Id not only stimulates responses in neonates, but also, by recruiting the T-cell system, may stimulate immunological memory.

A second situation where anti-Id, but not antigen, stimulates a particular antibody response is in the activation of silent clones. Following immunization of BALB/c mice with BL, the A48 Id is not normally expressed, but mice primed at birth with anti-A48 do express A48 in response to a subsequent immunization with BL.[68,69] The ability to selectively stimulate a particular Id is an important aspect of the use of anti-Id vaccines, inasmuch as antibodies of different Id that bind the same antigen have been shown to have different protective capacities against infection.[70,71]

Finally, anti-Id may be useful in stimulating anti-PS responses in individuals with immunodeficiencies that render them unresponsive to PS antigens. That thymus-dependent conjugates could stimulate an anti-PS response was demonstrated in mice with the *xid* defect,[72] and recently this has been confirmed in a child with an IgG2 deficiency.[73] Recent experiments have shown that anti-Id can also prime *xid* mice for a subsequent response to PS.[90]

D. Anti-Id Can Be Used When Antigen Is Not Available or Is Harmful

There are several situations in which antigen is unavailable in a form suitable for a vaccine. One situation is when the antigen is present in the organism in a very low concentration. The purification of such antigens in large quantities and of sufficient purity is very difficult. The techniques of recombinant DNA technology ultimately may solve this problem for protein antigens that are the products of a single gene. Another problem for conventional vaccine production concerns antigens that are toxic. Even when toxins can be prepared in large quantities, they must be detoxified sufficiently that they do not revert, yet they maintain their antigenicity.

Bacteria that multiply within cells present another type of problem. First, T-cell immunity is required for protection.[4] Second, killed whole cell vaccines or products of these cells are usually not as effective as live vaccines in stimulating T-cell immunity.[32,74] One possible approach is the use of an anti-Id directed at the T-cell receptor for antigen.[75] T-cell clones specific for protective antigens have been used to prepare anticlonotypic antibodies and have been shown to induce protective T-cell immunity against a viral infection,[76] and as will be discussed, against a bacterial infection.[77]

Table 2
CRITERIA FOR ANTI-ID SELECTION FOR
VACCINES

The Id (in the form of antibody or T cells) should confer passive
immunity
The anti-Id-Id interaction should be inhibited by antigen
The anti-Id should induce Id regardless of the allotype of the
recipient

III. ANTI-ID AS BACTERIAL VACCINES

A. Introduction

The criteria for selection of an anti-Id to be used as a vaccine are outlined in Table 2. The goal of any immunization is protective immunity. Thus, the most important criterion for the selection of an anti-Id vaccine is that the Id itself (either serum antibody or T cells) be protective. Before an Id is selected for use in the production of an anti-Id, its ability to confer passive protection in an appropriate animal model or in vitro bactericidal assay should be demonstrated. In the anti-Id-treated host, the Id should also be shown to increase after exposure to antigen. This is important in order to ensure that the response will be amplified upon infection with the organism. The anti-Id should be specific for the antigen combining site of the Id in order to maximize the chances of inducing antigen-binding Id (as opposed to Id-positive molecules that do not bind antigen). Finally, the ideal anti-Id vaccine should stimulate Id production in a genetically unrestricted fashion so that it may be used in outbred populations. It has been suggested by Nisonoff and colleagues[40,78] that a requirement for an anti-Id vaccine is that it possess an internal image of antigen, i.e., be an $Ab_{2\beta}$ (in their terms, "related epitope") in order to be a useful vaccine. An $Ab_{2\alpha}$ that, via its paratope (combining site), reacts with a conserved determinant or structure on antibodies from individuals of different allotypes or different species, however, would also qualify as a vaccine candidate. To date, there are no structural data available to distinguish between these possibilities for the anti-Id that have been used as bacterial vaccines. Much of the work on anti-Id vaccines concerns the use of viral pathogens; these will be reviewed elsewhere in these volumes. Below, I will review the bacterial systems that have used anti-Id to prime for or induce protective immunity.

B. *Escherichia coli* K13

The presence of antibodies directed against the capsular PS antigens of pathogenic bacteria has been shown to correlate with protection from infection caused by these organisms.[1-3] PS antigens, however, fail to stimulate antibody responses in neonates.[2,15-18] Newborn mice also fail to make anti-PS antibodies even when immunized with oligosaccharide-protein conjugates.[72] Because of these findings and the obvious desirability of stimulating anti-PS antibodies early in life, a time of increased incidence of invasive disease due to encapsulated organisms, Soderstrom and I studied the effects of network manipulation during the neonatal period on the subsequent ability to immunize for protective immunity.[79] The model system chosen was that of *E. coli* K13, an encapsulated human pathogen.[80,81] This organism kills mice in 24 hr when injected i.p. and protection can be achieved by the administration of passive antibody directed at the capsule.[82] The Id used in our experiments was 150C8, an IgM monoclonal antibody from BALB/c mice that is specific for the capsular K13 PS. The anti-Id was 5868C, an IgG1 monoclonal antibody from A/HeJ mice. We showed that 150C8 can confer passive protection in an animal model[82] and is bactericidal in vitro.[83] In addition, 150C8 Id appears in the sera of adult mice of different allotypes and in outbred rats following immunization with K13 PS or killed cells.[79] The 5868C-150C8 interaction was demonstrated

Table 3
EFFECTS OF ANTI-ID PRIMING ON SURVIVAL FOLLOWING A LETHAL CHALLENGE WITH *E. COLI* K13

Treatment at birth	Age at immuni- zation (weeks)	Antigen[a]	Age at challenge (weeks)	Challenge dose[b] (LD_{50})	No. of mice	Survival (%)
— [c]	4	—	5	20	6	0
K13 PS[d]	4	K13 vaccine[e]	5	20	6	33
Anti-150C8[f]	4	K13 vaccine	5	20	15	87
Saline	4	K13 PS	5	30	6	0
K13 PS	4	K13 PS	5	30	6	33
Anti-150C8	4	K13 PS	5	30	9	78
—	12	—	13	50	9	0
—	12	K13 PS	13	50	20	25
Anti-150C8	12	K13 PS	13	50	30	78

[a]   Mice were immunized wtih antigen in phosphate-buffered saline (PBS) by i.p. injection.
[b]   *E. coli* O6:K13:H1 in PBS was used for challenge by i.p. injection. One LD_{50} is approximately 5 × 10^6 bacteria.
[c]   No injection.
[d]   2.5 μg K13 PS.
[e]   1 × 10^8 Killed K13 bacteria in PBS.
[f]   Anti-150C8 is a monoclonal IgG1 antibody, 5868C, specific for the IgM monoclonal, 150C8, that binds the K13 capsular PS; 50 ng was given in PBS by i.p. injection.

From Stein, K. E. and Soderstrom, T., *J. Exp. Med.*, 160, 1001, 1984. With permission.

to be specifically inhibited by K13 PS.[79] Most important, these studies demonstrated that mice treated at birth with a low dose of anti-Id, but not K13 PS itself, were primed for a subsequent protective response following immunization with K13 and challenge with a lethal dose of bacteria.[79] These results, some of which are shown in Table 3, also demonstrated that not only weanling mice (immunized at 4 weeks of age), but also adult mice (immunized at 12 weeks of age) exhibited enhanced protection as a result of neonatal priming. Thus, it can be seen that the administration of as little as 50 ng of monoclonal anti-Id at birth can make the difference between mice living and dying following a lethal challenge.

C. *Streptococcus pneumoniae*
An anti-Id vaccine has been used to induce protective immunity against Type 3 *S. pneumoniae* in adult mice.[84,85] In these experiments the Id, T15, was a BALB/c IgA myeloma protein specific for the phosphoryl choline (PC) hapten on the cell wall of *S. pneumoniae*, and the anti-Id, 4C11, was an A/He anti-T15. The anti-Id was coupled to keyhole limpet hemocyanin (KLH) and the conjugate, 4C11-KLH, was used to immunize mice prior to challenge with *S. pneumoniae*. PC-KLH was used as a positive control. Mice were immunized with 4C11-KLH or control proteins in complete Freund adjuvant on two occasions and challenged with a lethal dose of bacteria 2 weeks after the last injection. The experiments showed that 4C11-KLH was as effective in stimulating anti-PC and in achieving protection against Type 3 *S. pneumoniae* in BALB/c mice as PC-KLH.[84] Interestingly, when A/He mice were immunized with 4C11-KLH, they produced anti-PC antibodies of the T15 Id, however, no protection was seen.[85] The reason for the lack of efficacy in A/He mice is unknown.

D. Bacterial Toxins
Although bacterial toxins are ideal candidates for the development of anti-Id vaccines, I

am not aware of any reports of such vaccines. Nonetheless, there are some data which indicate that this approach will work. Lipopolysaccharide (LPS), bacterial endotoxin, plays a role in the pathogenesis of Gram-negative sepsis and antibodies to LPS have been reported to be protective when passively administered to patients with septicemia.[86,87] Hiernaux and Bona have reported the existence of shared Id on monoclonal antibodies specific for the LPS of different Gram-negative bacteria.[88] A human monoclonal antibody specific for LPS was described and shown to protect mice from Gram-negative septicemia.[89] Thus, an anti-Id vaccine for stimulating protective anti-LPS in man should be possible. Recently, Sieckmann has reported[91] that an anti-Id made against anti-exotoxin A of *Pseudomonas aeruginosa* can prime mice for a two- to eightfold increased antibody response to a subsequent immunization with exotoxin A toxoid compared to mice not primed with anti-Id. It may be possible, therefore, that anti-Id priming for this type of toxin would permit the use of very small doses of toxoid for immunization, thereby minimizing the possible side effects of a toxoid vaccine.

E. *Listeria monocytogenes*

As discussed above, the fact that T-cell immunity is required for protection against some bacteria, specifically those that are facultative intracellular parasites,[4] does not preclude the use of an anti-Id vaccine. Kaufmann et al.[77] prepared anticlonotypic sera against a T-cell hybridoma specific for *L. monocytogenes* strain EGD. They demonstrated that the T-cell hybrid, when injected together with the bacteria, protected mice from infection and that the hybrid recognizes antigen in association with Class II antigens of the major histocompatibility complex (MHC). When anti-Id sera were used to immunize mice prior to challenge they were shown to protect the animals from listeriosis in an antigen-specific manner. An important aspect of the protection induced by anti-Id is that it was not limited to particular alleles of the MHC complex or the immunoglobulin locus. This is an obviously desirable feature of an anti-Id vaccine to be used in outbred populations.

IV. CONCLUSIONS

In this chapter I have attempted to demonstrate that anti-Id vaccines have a place in the armamentarium of bacterial vaccines. Although in most instances anti-Id vaccines will not replace conventional types, there are many situations in which conventional vaccines do not exist, are harmful, or are not effective. For these situations, anti-Id vaccines, if appropriately selected and prepared, offer promise. The fact that monoclonal anti-Id antibodies can function as vaccines offers the possibility of producing large quantities of well-characterized, efficacious, and safe vaccines at a reasonable cost.

REFERENCES

1. **Kaijser, B. and Ahlstedt, S.**, Protective capacity of antibodies against *Escherichia coli* O and K antigens, *Infect. Immun.*, 17, 286, 1977.
2. **Schneerson, R., Rodrigues, L. P., Parke, J. C., Jr., and Robbins, J. B.**, Immunity to disease caused by *Hemophilus influenzae* type b. II. Specificity and some biologic characteristics of "natural", infection-acquired, and immunization-induced antibodies to the capsular polysaccharide of *Hemophilus influenzae* type b, *J. Immunol.*, 107, 1081, 1971.
3. **Egan, M. L., Pritchard, D. G., Dillon, H. C., Jr., and Gray, B. M.**, Protection of mice from experimental infection with type III group B streptococcus using monoclonal antibodies, *J. Exp. Med.*, 158, 1006, 1983.
4. **Hahn, H. and Kaufmann, H. E.**, The role of cell-mediated immunity in bacterial infections, *Rev. Infect. Dis.*, 3, 1221, 1981.

5. **Germanier, R., Ed.,** *Bacterial Vaccines,* Academic Press, New York, 1984.
6. **Bizzini, B.,** Tetanus, in *Bacterial Vaccines,* Germanier, R., Ed., Academic Press, New York, 1984, 38.
7. **Pappenheimer, A. M., Jr.,** Diphtheria, in *Bacterial Vaccines,* Germanier, R., Ed., Academic Press, New York, 1984, 1.
8. **Manclark, C. R. and Cowell, J. L.,** Pertussis, in *Bacterial Vaccines,* Germanier, R., Ed., Academic Press, New York, 1984, 69.
9. **Finkelstein, R. A.,** Cholera, in *Bacterial Vaccines,* Germanier, R., Ed., Academic Press, New York, 1984, 107.
10. **Sato, Y., Kimura, M., and Fukumi, H.,** Development of a pertussis component vaccine in Japan, *Lancet,* i, 122, 1984.
11. **Band, J. D., Chamberland, M. E., Platt, T., Weaver, R. E., Thornsberry, C., and Fraser, D. W.,** Trends in meningococcal disease in the United States, 1975-1980, *J. Infect. Dis.,* 148, 754, 1983.
12. **Kayhty, H., Karanko, V., Peltola, H., and Makela, P. H.,** Serum antibodies after vaccination with *Haemophilus influenzae* type b capsular polysaccharide and responses to reimmunization: no evidence of immunologic tolerance or memory, *Pediatrics,* 74, 857, 1984.
13. **Makela, P. H., Sibakov, M., Herva, E., Henrichsen, J., Luotonen, J., Timonen, J., Leinonen, M., Koskela, M., Pukander, J., Pontynen, S., Gronroos, P., and Karma, P.,** Pneumococcal vaccine and otitis media, *Lancet,* ii, 547, 1980.
14. **Robbins, J. B., Austrian, R., Lee, C.-J., Rastogi, S. C., Schiffman, G., Henrichsen, J., Makela, P. H., Broome, C. V., Facklam, R. R., Tiesjema, R. H., and Parke, J. C., Jr.,** Considerations for formulating the second-generation pneumococcal capsular polysaccharide vaccine with emphasis on the cross-reactive types within groups, *J. Infect. Dis.,* 148, 1136, 1983.
15. **Smith, D. H., Peter, G., Ingram, D. L., Harding, A. L., and Anderson, P.,** Responses of children immunized with the capsular polysaccharide of *Hemophilus influenzae* type b, *Pediatrics,* 52, 637, 1973.
16. **Peltola, H., Kayhty, H., Sivonen, A., and Makela, P. H.,** *Haemophilus influenzae* in children: a double-blind field study of 100,000 vaccinees 3 months to 5 years of age in Finland, *Pediatrics,* 60, 730, 1977.
17. **Pincus, D. J., Morrison, D., Andrews, C., Lawrence, E., Sell, S. H., and Wright, P. F.,** Age-related response to two *Haemophilus influenzae* type b vaccines, *J. Pediatr.,* 100, 197, 1982.
18. **Parke, J. C., Jr., Schneerson, R., Robbins, J. B. and Schlesselman, J. J.,** Interim report of a controlled field trial immunization with capsular polysaccharides of *Haemophilus influenzae* type b and group C *Neisseria meningitidis* in Mecklenburg County, North Carolina (March 1974—March 1976, *J. Infect. Dis.,* 136, S51, 1977.
19. **Chu, C., Schneerson, R., Robbins, J. B., and Rastogi, S. C.,** Further studies on the immunogenicity of *Haemophilus influenzae* type b and penumococcal type 6A polysaccharide-protein conjugates, *Infect. Immun.,* 40, 245, 1983.
20. **Beuvery, E. C., van Delft, R. W., Miedema, F., Kanhai, V., and Nagel, J.,** Immunological evaluation of meningococcal group C polysaccharide-tetanus toxoid conjugate in mice, *Infect. Immun.,* 41, 609, 1983.
21. **Schneerson, R., Robbins, J. B., Chu, C., Sutton, A., Vann, W., Vickers, J. C., London, W. T., Curfman, B., Hardegree, M. C., Shiloach, J., and Rastogi, S. C.,** Serum antibody response of juvenile and infant rhesus monkeys injected with *Haemophilus influenzae* type b and pneumococcus type 6A capsular polysaccharide-protein conjugates, *Infect. Immun.,* 45, 582, 1984.
22. **Granoff, D. M., Boies, E. G., and Munson, R. S., Jr.,** Immunogenicity of *Haemophilus influenzae* type b polysaccharide-diphtheria toxoid conjugate vaccine in adults, *J. Pediatr.,* 105, 22, 1984.
23. **Eskola, J., Kayhty, H., Peltola, H., Karanko, V., Makela, P. H., Samuelson, J., and Gordon, L. K.,** Antibody levels achieved in infants by course of *Haemophilus influenzae* type b polysaccharide/diphtheria toxoid conjugate vaccine, *Lancet,* i, 1184, 1985.
24. **Jorbeck, H. J. A., Svenson, S. B., and Lindberg, A. A.,** Artificial *Salmonella* vaccines: *Salmonella typhimurium* O-antigen-specific oligosaccharide-protein conjugates elicit opsonizing antibodies that enhance phagocytosis, *Infect. Immun.,* 32, 497, 1981.
25. **Anderson, P.,** Antibody responses to *Haemophilus influenzae* type b and diphtheria toxin induced by conjugates of oligosaccharides of the type b capsule with the nontoxic protein CRM_{197}, *Infect. Immun.,* 39, 233, 1983.
26. **Jennings, H. J., Lugowski, C., and Ashton, F. E.,** Conjugation of meningococcal lipopolysaccharide R–type oligosaccharides to tetanus toxoid as route to a potential vaccine against group B *Neisseria meningitidis, Infect. Immun.,* 43, 407, 1984.
27. **Stein, K. E., Zopf, D. A., Johnson, B. M., Miller, C. B., and Paul, W. E.,** The immune response to an isomaltohexosyl-protein conjugate, a thymus-dependent analogue of $\alpha(1{\rightarrow}6)$ dextran, *J. Immunol.,* 128, 1350, 1982.
28. **Anderson, P., Pichichero, M. E., and Insel, R. A.,** Immunization of 2-month-old infants with protein-coupled oligosaccharides derived from the capsule of *Haemophilus influenzae* type b, *J. Pediatr.,* 107, 346, 1985.

29. **Lepow, M. L., Samuelson, J. S., and Gordon, L. K.,** Safety and immunogenicity of *Haemophilus influenzae* type b-polysaccharide diphtheria toxoid conjugate vaccine in infants 9 to 15 months of age, *J. Pediatr.*, 106, 185, 1985.

30. **Cates, K. L.,** Serum opsonic activity for *Haemophilus influenzae* type b in infants immunized with polysacchride-protein conjugate vaccines, *J. Infect. Dis.*, 152, 1076, 1985.

31. **Insel, R. A. and Anderson, P. W.,** Oligosaccharide-protein conjugate vaccines induce and prime for oligoclonal IgG antibody responses to the *Haemophilus influenzae* b capsular polysaccharide in human infants, *J. Exp. Med.*, 163, 262, 1986.

32. **Collins, F. M.,** Tuberculosis, in *Bacterial Vaccines*, Germanier, R., Ed., Academic Press, New York, 1984, 373.

33. **Godal, T.,** Leprosy, in *Bacterial Vaccines*, Germanier, R., Ed., Academic Press, New York, 1984, 419.

34. **Allen, P. M., Beller, D. I., Braun, J., and Unanue, E. R.,** The handling of *Listeria monocytogenes* by macrophages: the search for an immunogenic molecule in antigen presentation, *J. Immunol.*, 132, 323, 1984.

35. **Germanier, R.,** Thyphoid fever, in *Bacterial Vaccines*, Germanier, R., Ed., Academic Press, New York, 1984, 137.

36. **Blanden, R. V., Mackaness, G. B., and Collins, F. M.,** Mechanisms of acquired resistance in mouse typhoid, *J. Exp. Med.*, 124, 585, 1966.

37. **Hochadel, J. F. and Keller, K. F.,** Protective effects of passively transferred immune T- or B-lymphocytes in mice infected with *Salmonella typhimurium*, *J. Infect. Dis.*, 135, 813, 1977.

38. **Jerne, N. K., Roland, J., and Cazenave, P.-A.,** Recurrent idiotopes and internal images, *EMBO J.*, 1, 243, 1982.

39. **Jerne, N. K.,** Towards a network theory of the immune system, *Ann. Immunol. (Inst. Pasteur)*, 125 C, 373, 1974.

40. **Nisonoff, A. and Lamoyi, E.,** Hypothesis implications of the presence of an internal image of the antigen in anti-idiotypic antibodies: possible application to vaccine production, *Clin. Immunol. Immunopathol.*, 21, 397, 1981.

41. **Bona, C., Lieberman, R., Chien, C. C., Mond, J., House, S., Green, I., and Paul, W. E.,** Immune response to levan. I. Kinetics and ontogeny of anti-levan and anti-inulin antibody response and of expression of cross-reactive idiotype, *J. Immunol.*, 120, 1436, 1978.

42. **Fernandez, C. and Moller, G.,** Antigen-induced strain-specific autoantiidiotypic antibodies modulate the immune response to dextran B 512, *Proc. Natl. Acad. Sci. U.S.A.*, 76, 5944, 1979.

43. **Fernandez, C. and Moller, G.,** A primary immune response to dextran B512 is followed by a period of antigen-specific immunosuppression caused by autoanti-idiotypic antibodies, *Scand. J. Immunol.*, 11, 53, 1980.

44. **Rodkey, L. S., Binion, S. B., Brown, J. C., and Adler, F. L.,** Natural regulation of antibody clones by auto-anti-idiotypic antibodies in rabbits, in *Immune Networks*, Bona, C. A. and Kohler, H., Eds., *Ann. N.Y. Acad. Sci.*, 418, 16, 1983.

45. **Schrater, A. F., Goidl, E. A., Thorbecke, G. J., and Siskind, G. W.,** Production of atuo-anti-idiotypic antibody during the normal immune response to TNP-ficoll. I. Occurrence in AKR/J and BALB/c mice to hapten-augmentable, anti-TNP plaque-forming cells and their accelerated appearance in recipients of immune spleen cells, *J. Exp. Med.*, 150, 138, 1979.

46. **Geha, R. S.,** Presence of auto-anti-idiotypic antibody during the normal human immune response to tetanus toxoid antigen, *J. Immunol.*, 129, 139, 1982.

47. **Geha, R. S.,** Presence of circulating anti-idiotype-bearing cells after booster immunization with tetanus toxoid (TT) and inhibition of anti-TT antibody synthesis by auto-anti-idiotypic antibody, *J. Immunol.*, 130, 1634, 1983.

48. **Saxon, A. and Barnett, E.,** Human auto-antiidiotypes regulating T cell-mediated reactivity to tetanus toxoid, *J. Clin. Invest.*, 73, 342, 1984.

49. **Briles, D. E. and Davie, J. M.,** Clonal dominance. I. Restricted nature of the IgM antibody response to group A streptococcal carbohydrate in mice, *J. Exp. Med.*, 141, 1291, 1975.

50. **Hansburg, D., Briles, D. E., and Davie, J. M.,** Analysis of the diversity of murine antibodies to dextran B1355. I. Generation of a large, pauci-clonal response by a bacterial vaccine, *J. Immunol.*, 117, 569, 1976.

51. **Hansburg, D., Clevinger, B., Perlmutter, R. M., Griffith, R., Briles, D. E., and Davie, J. M.,** Analysis of the diversity of murine antibodies to $\alpha(1\rightarrow3)$ dextran, in *Cells of Immunoglobulin Synthesis*, Academic Press, New York, 1979, 295.

52. **Stein, K. E., Bona, C., Lieberman, R., Chien, C. C., and Paul, W. E.,** Regulation of the anti-inulin antibody response by a nonallotype-linked gene, *J. Exp. Med.*, 151, 1088, 1980.

53. **Wikler, M. and Urbain, J.,** Idiotypic manipulation of the rabbit immune response against *Micrococcus luteus*, in *Idiotypy in Biology and Medicine*, Kohler, H., Urbain, J., and Cazenave, P.-A., Eds., Academic Press, New York, 1984, 219.

54. **Kunkel, H. G., Mannik, M., and Williams, R. C.,** Individual antigenic specificity of isolated antibodies, *Science,* 140, 1218, 1963.

55. **Yount, W. J., Dorner, M. M., Kunkel, H. G., and Kabat, E. A.,** Studies on human antibodies. VI. Selective variations in subgroup composition and genetic markers, *J. Exp. Med.,* 127, 633, 1968.

56. **Insel, R. A., Kittelberger, A., and Anderson, P.,** Isoelectric focusing of human antibody to the *Haemophilus influenzae* b capsular polysaccharide: restricted and identical spectrotypes in adults, *J. Immunol.,* 135, 2810, 1985.

57. **Altevogt, P. and Wigzell, H.,** A V_H-associated idiotype in human anti-tetanus antibodies, *Scand. J. Immunol.,* 17, 183, 1983.

58. **Hoffman, W. L., Strucely, P. D., Jump, A. A., and Smiley, J. D.,** A restricted human antitetanus clonotype shares idiotypic cross-reactivity with tetanus antibodies from most human donors and rabbits: reactivity with antibodies of widely differing electrophorectic mobility, *J. Immunol.,* 135, 3802, 1985.

59. **Morrison, S. L., Johnson, M. J., Herzenberg, L. A., and Oi, V. T.,** Chimeric human antibody molecules: mouse antigen-binding domains with human constant region domains, *Proc. Natl. Acad. Sci. U.S.A.,* 81, 6851, 1984.

60. **Bona, C., Mond, J. J., Stein, K. E., House, S., Lieberman, R., and Paul, W. E.,** Immune response to levan. III. The capacity to produce anti-inulin antibodies and cross-reactive idiotypes appears late in ontogeny, *J. Immunol.,* 123, 1484, 1979.

61. **Mosier, D. E., Zaldivar, N. M., Goldings, E., Mond, J., Sher, I., and Paul, W. E.,** Formation of antibody in the newborn mouse: study of T-cell-independent antibody response, *J. Infect. Dis.,* 136, S14, 1977.

62. **Baker, P. J., Morse, H. C., III, Cross, S. S., Stashak, P. W., and Prescott, B.,** Maturation of regulatory factors influencing magnitude of antibody response to capsular polysaccharide of type III *Streptococcus pneumoniae, J. Infect. Dis.,* 136, S20, 1977.

63. **Mosier, D. E., Zitron, I. M., Mond, J. J., Ahmed, A., Sher, I., and Paul, W. E.,** Surface immunoglobulin D as a functional receptor for a subclass of B lymphocytes, *Immunol. Rev.,* 37, 89, 1977.

64. **Gold, R., Lepow, M. L., Goldschneider, I., and Gotschlich, E. C.,** Immune response of human infants to polysaccharide vaccines of groups A and C *Neisseria meningitidis, J. Infect. Dis.,* 136, S31, 1977.

65. **Miller, G. G., Nadler, P. I., Asano, Y., Hodes, R. J., and Sachs, D. H.,** Induction of idiotype-bearing, nuclease-specific helper T cells by in vivo treatment with anti-idiotype, *J. Exp. Med.,* 154, 24, 1981.

66. **Heidelberger, M.,** Persistence of antibodies in man after immunization, in *The Nature and Significance of the Antibody Response,* Pappenheimer, A. M., Jr., Ed., Columbia University Press, New York, 1953, chap. 5.

67. **Anderson, P., Peter, G., Johnston, R. B., Jr., Wetterlow, L. H., and Smith, D. H.,** Immunization of humans with polyribophosphate, the capsular antigen of *Hemophilus influenzae,* type b, *J. Clin. Invest.,* 51, 39, 1972.

68. **Hiernaux, J., Bona, C., and Baker, P. J.,** Neonatal treatment with low doses of anti-idiotypic antibody leads to the expression of a silent clone, *J. Exp. Med.,* 153, 1004, 1981.

69. **Rubinstein, L. J., Goldberg, B., Hiernaux, J., Stein, K. E., and Bona, C. A.,** Idiotype-antiidiotype regulation. V. The requirement for immunization with antigen or monoclonal antiidiotypic antibodies for the activation of β2→6 and β2→1 polyfructosan-reactive clones in BALB/c mice treated at birth with minute amounts of anti-A48 idiotype antibodies, *J. Exp. Med.,* 158, 1129, 1983.

70. **Briles, D. E., Forman, C., Hudak, S., and Claflin, J. L.,** Anti-phosphorylcholine antibodies of the T15 idiotype are optimally protective against *Streptococcus pneumoniae, J. Exp. Med.,* 156, 1177, 1982.

71. **Briles, D. E., Forman, C., Hudak, S., and Claflin, J. L.,** The effects of idiotype on the ability of IgG₁ anti-phosphocholine antibodies to protect mice from fatal infection with *Streptococcus pneumoniae, Eur. J. Immunol.,* 14, 1027, 1984.

72. **Stein, K. E., Zopf, D. A., Miller, C. B., Johnson, B. M., Mongini, P. K. A., Ahmed, A., and Paul, W. E.,** Immune response to a thymus-dependent form of B512 dextra requires the presence of Lyb-5⁺ lymphocytes, *J. Exp. Med.,* 157, 657, 1983.

73. **Insel, R. A. and Anderson, P. W.,** Bypass of antibody unresponsiveness to a bacterial capsular polysaccharide in IgG2 subclass deficiency by using an oligosaccharide conjugate vaccine, *N. Engl. J. Med.,* 315, 499, 1986.

74. **Tomita, T., Blumenstock, E., and Kanegasaki, S.,** Phagoctic and chemiluminescent responses of mouse peritoneal macrophages to living and killed *Salmonella typhimurium* and other bacteria, *Infect. Immun.,* 32, 1242, 1981.

75. **Haskins, K., Kappler, J., and Marrack, P.,** The major histocompatibility complex-restricted antigen receptor on T cells, *Annu. Rev. Immunol.,* 2, 51, 1984.

76. **Ertl, H. C. J. and Finberg, R. W.,** Sendai virus-specific T-cell clones: induction of cytolytic T cells by an anti-idiotypic antibody directed against a helper T-cell clone, *Proc. Natl. Acad. Sci. U.S.A.,* 81, 2850, 1984.

77. **Kaufmann, S. H. E., Eichmann, K., Muller, I., and Wrazel, L. J.,** Vaccination against the intracellular bacterium *Listeria monocytogenes* with a clonotypic antiserum, *J. Immunol.,* 134, 4123, 1985.
78. **Gourish, M. F. and Nisonoff, A.,** Potential use of anti-idiotype antibodies as vaccines, in *Proc. 1st Annual Southwest Foundation for Biomedical Research Int. Symp.,* Houston, Texas, Nov. 8—10, 1984, Dreesman, G. R., Bronson, J. G., and Kennedy, R. C., Eds., American Society of Microbiology, Washington, D.C., 1985, 103.
79. **Stein, K. E. and Soderstrom, T.,** Neonatal administration of idiotype or antiidiotype primes for protection against *Escherichia coli* K13 infection in mice, *J. Exp. Med.,* 160, 1001, 1984.
80. **Kaijser, B., Hanson, L. A., Jodal, U., Lindin-Janson, G., and Robbins, J. B.,** Frequency of *E. coli* K antigens in urinary tract infections in children, *Lancet,* i, 663, 1977.
81. **Vann, W. F., Soderstrom, T., Egan, W., Tsui, F.-P., Schneerson, R., Orskov, I., and Orskov, F.,** Serological, chemical, and structural analyses of the *Escherichia coli* cross-reactive capsular polysaccharides K13, K20, and K23, *Infect. Immun.,* 39, 623, 1983.
82. **Soderstrom, T., Brinton, C. C., Jr., Fusco, P., Karpas, A., Ahlstedt, S., Stein, K., Sutton, A., Hosea, S., Schneerson, R., and Hanson, L. A.,** Analysis of pilus-mediated pathogenic mechanism with monoclonal antibodies, in *Microbiology—1982,* Schlessinger, D., Ed., American Society of Microbiology, Washington, D. C., 1982, 305.
83. **Soderstrom, T., Stein, K., Brinton, C. C., Jr., Hosea, S., Burch, C., Hansson, H. A., Karpas, A., Schneerson, R., Sutton, A., Vann, W. F., and Hanson, L. A.,** Serological and functional properties of monoclonal antibodies to *Escherichia coli* type I pilus and capsular antigens, *Prog. Allergy,* 33, 259, 1983.
84. **MacNamara, M. K., Ward, R. E., and Kohler, H.,** Monoclonal idiotope vaccine against *Streptococcus pneumoniae* infection, *Science,* 226, 1325, 1984.
85. **Kohler, H., McNamara, M., and Ward, R.,** Monoclonal idiotype vaccines, in *Proc. 1st Annual Southwest Foundation for Biomedical Research Int. Symp.,* Houston, Texas, Nov. 8—10, 1984, Dreesman, G. R., Bronson, J. G., and Kennedy, R. C., Eds. American Society of Microbiology, Washington, D.C., 1985, 161.
86. **Ziegler, E. J., McCutchan, J. A., Fierer, J., Glauser, M. P., Sadoff, J. C., Douglas, H., and Braude, A. I.,** Treatment of Gram-negative bacteremia and shock with human antiserum to a mutant *Escherichia coli, N.Engl. J. Med.,* 307, 1225, 1982.
87. **Baumgartner, J.-D., Glauser, M. P., McCutchan, J. A., Ziegler, E. J., Melle, G., Klauber, M. R., Vogt, M., Muehlen, E., Luethy, R., Chiolero, R., and Geroulanos, S.,** Prevention of Gram-negative shock and death in surgical patients by antibody to endotoxin core glycolipid, *Lancet,* ii, 59, 1985.
88. **Hiernaux, J. and Bona, C. A.,** Shared idiotypes among monoclonal antibodies specific for different immunodominant sugars of lipopolysaccharide of different Gram-negative bacteria, *Proc. Natl. Acad. Sci. U.S.A.,* 79, 1616, 1982.
89. **Teng, N. N. H., Kaplan, H. S., Herbert, J. M., Moore, C., Douglas, H., Wunderlich, A., and Braude, A. I.,** Protection against Gram-negative bacteremia and entotoxemia with human monoclonal IgM antibodies, *Proc. Natl. Acad. Sci. U.S.A.,* 82, 1790, 1985.
90. **Stein, K. E.,** Restoration of the ability to respond to polysaccharide (PS) antigens in *xid* mice by neonatal priming with Id or anti-Id, *Fed. Proc. Fed. Am. Soc. Exp. Biol.,* 46, 1024, 1987.
91. **Sieckmann, D.,** personal communication.

Chapter 2

MODULATION OF THE IMMUNE SYSTEM TOWARDS ANTIVIRAL IMMUNE RESPONSE USING ANTI-IDIOTYPIC STRUCTURES

A. D. M. E. Osterhaus, E. J. Bunschoten, K. Weijer, and F. G. C. M. UytdeHaag

TABLE OF CONTENTS

I. INTRODUCTION

The control of virus diseases still depends on epidemiologic measures including the exertion of vaccination strategies. For many virus infections, vaccination has proven to be quite effective and in the case of smallpox even resulted in the total eradication of the disease, about two centuries after the initial introduction of "vaccination" by Jenner in 1798. Conventional vaccines based upon whole live-attenuated or inactivated viruses may, however, still present major complications (for review see Reference 1). Attenuated virus strains may reverse to virulence, may be too pathogenic for immune-deficient individuals, and cannot be used in the presence of maternal antibody. Vaccination with inactivated viruses requires adequate systems for large-scale production and purification to ensure the absence of contaminating nucleic acids and proteins which may cause adverse reactions. The use of these inactivated vaccines, but also of subunit vaccines, which only include those components of the virus which elicit the desired immune response, is greatly hampered by the lack of adequate systems for their immunogenic presentation. With the advent of recombinant DNA technology and the possibilities for chemical synthesis of peptides of defined amino acid sequences, it would seem attractive to only use well-defined immunogenic moieties as inert vaccines. However, it has been shown that compared to live viruses, also these molecules, even in the presence of adjuvants, generally induce inferior immune responses and do not mediate long-lasting protection. Although the reasons are not fully understood, at present we know that during virus infection, viral antigens are expressed in large amounts in association with antigens of the major histocompatibility complex (MHC) on cell membranes and, consequently, a vigorous T-cell-mediated immune response is induced. Protection against many acute virus infections seems to depend on T-cell-mediated immune mechanisms. Therefore, the novel approach of using genetically engineered recombinant viruses as live vaccines (e.g., engineered vaccinia strains) expressing genes encoding proteins of viral pathogens seems attractive (for review see Reference 2). The major problem with such vaccines, especially for application in humans, is generally believed to be the selection of appropriate viral vectors. This will depend upon the evaluation of their pathogenicity in laboratory animal systems. Moreover, these vaccines may, in principle, present similar complications as mentioned for conventional live-attenuated virus vaccines.

Even with these considerations taken together, a final and universal solution for the generation of safe and effective virus vaccines has not yet been found for animal viral diseases. Moreover, new viruses with different genome strategies and pathogenic mechanisms emerge from time to time in different animal species, canine parvovirus and the human T-lymphotropic retroviruses (HTLV I, II, and III) being recent examples. These reasons strongly motivate the search for novel approaches leading to alternative ways of vaccination against virus diseases.

Recent studies, based upon the network concept of the immune system initially proposed by Jerne,[3] have implicated the exploitation of elements of the immune system itself — anti-idiotype (anti-Id) antibodies — to replace "external" antigens for vaccination. The basis of this concept is the mutal recognition of variable (V) domains of the elements of the immune system which form a permanently communicating web of idiotypes (Id) and anti-Id. The amino acid sequence in the V region of an immunoglobulin determines the three-dimensional conformation of its antigen-combining site, which permits it to react with a certain epitope on a particular antigen or with a certain idiotope located within a particular Id. An antibody (Ab_1) response upon injection with an antigen, or the injection of an Ab_1 itself, results in an anti-Id (Ab_2) response recognizing idiotopes of the Ab_1, which, in turn, will elicit an Ab_3 response regulating this Ab_2 response by recognizing its Id specificities. Since an Ab_2 may react with the antigen-binding site within the V region of Ab_1 on the basis of a structural complementarity, it may represent a three-dimensional internal image of the epitope on the external antigen. Consequently, the administration of Ab_2 should be able to trigger the immune system to expand a population of Ab_3-producing cells which not only bind to Ab_2, but also bind to antigen and may, thus, be used as a substitute for the administration of antigen. In the present paper, we shall review data from various groups including our own (Table 1) which clearly show that Ab_2 in animal virus systems, indeed, are capable of inducing antiviral immunity and that this capacity is not restricted to Ab_2 displaying the properties of true internal images. Different strategies using xenogeneic or allogeneic polyclonal and monoclonal Ab_2 have been explored. The hybridoma technology introduced in 1975 by Kohler and Milstein,[4] which permits the generation and large-scale production of unlimited amounts of monoclonal Ab_2 of predefined Id specificity, may prove to become an essential tool for the practical use of Ab_2 as vaccines.

II. INTERNAL IMAGES; Ab_2 DEFINING CRI

The possibility that anti-Id antibody may express idiotopes which show similarity with antigenic epitopes was first considered by Lindenman[5] who used the term "homobodies", and Nisonoff and Lamoyi indicated that this similarity would not necessarily indicate identity and, therefore, used the term "related epitope" (RE).[6] They also proposed the possible application of these RE or internal images for vaccine production. Sege and Peterson showed that Ab_2 prepared against bovine anti-insulin antibodies exhibited biological activity similar to insulin. They indicated that this could be explained by assuming that at least part of the Ab_2 population resembled structurally an antigenic determinant present on the insulin molecule.[7] Nisonoff and Lamoyi[6] indicated the following criteria which provide strong evidence that a given Ab_2 may, indeed, be considered an internal image, although the presence or absence of none of these criteria should be considered conclusive: (1) elicitation of an Ab_1-like response upon inoculation with Ab_2 in phylogenetically distant species; (2) reactivity between Ab_2 and an antigen receptor which is little or not genetically related to Ab_1 (e.g., a membrane receptor); (3) preservation of the reactivity with an Ab_1 when the binding site of Ab_2 is disrupted by separating its H and L chains; (4) reactivity of Ab_2 with a large proportion of Ab_1 from various sources, particularly if Ab_1 proves to be very heterogeneous; (5) reactivity of Ab_2 with Ab_1 from different animal species (cross-reactive Id: CRI), which is inhibitable by antigen. Ab_2 elicited by an Ab_1 reacting with a particular antigen may recognize, through its Id, idiotopes on Ab_1 which do not function as a paratope for the antigenic epitope and may be associated either with the framework or with the hypervariable region of Ab_1 ($Ab_{2\alpha}$). These idiotopes may be irrelevant to the interaction between antigen and Ab_1. Another population of Ab_2 ($Ab_{2\beta}$) binds through its Id to idiotopes on Ab_1 which do function as a paratope for the epitope. These $Ab_{2\beta}$ exhibit a structural resemblance to the epitope and may, thus, represent internal images of epitopes on (external) antigen (Figure 1). The Ab_3 population elicited by $Ab_{2\alpha}$ may, apart from its reactivity with $Ab_{2\alpha}$, share idiotopes with Ab_1. Part of the Ab_3 population may even exhibit antigen binding properties. These Ab_3 have, e.g., been demonstrated upon subsequent immunization with antigen by showing a priming effect. The incidence of Ab_3 with antigen binding properties, however, will be much higher in response to $Ab_{2\beta}$ (for review see Reference 8). Although, obviously, $Ab_{2\beta}$ internal images are the best candidates for future anti-Id vaccines, since their effects will in most cases not be restricted by genetic polymorphisms, from theoretical consideration mentioned above, it may be concluded that, also, Ab_2 not fulfilling $Ab_{2\beta}$ criteria may be considered as potential vaccines, and that problems with genetic restriction might be overcome. This was first demonstrated in a viral system by Francotte and Urbain who showed that conventional polyclonal rabbit Ab_2, raised against a *private* rabbit antitobacco mosaic virus (TMV) Id after coupling to lipopolysaccharide (LPS), induced an anti-TMV antibody response in another species (BALB/c mice) which strongly cross reacted with the starting

Table 1
MODULATION OF SPECIFIC IMMUNITY TO VIRUSES WITH Ab$_2$

Virus	Ab$_1$	Ab$_2$	Induction of	Ref.
Tobacco Mosaic Virus	Rabbit	Rabbit	Silent Id Ag-positive Ab$_3$ response	9
Venezuelan equine encephalitis	MoAb (α-GP56)	Rabbit (paratope specific)	Ag-positive Ab$_3$ response	50
Hepatitis B	Human (α-HBsAg)	Rabbit (anti-CRI, paratope specific)	Ag-positive Ab$_3$ response in mice, rabbits, chimpanzees (booster after HBsAg exposure	15 — 18
	MoAb (α-HBsAg)	MoAb (anti-CRI)		19
Poliovirus II	MoAb (VN)	MoAb (anti-CRI, paratope specific)	Ag-positive Ab$_3$ response (VN antibody)	8, 20, 21
Rabies virus	MoAb (VN)	Rabbit (paratope specific)	Ag-positive Ab$_3$ response (VN antibody) Booster effect Modulation of protection	30, 31
	MoAb (VN)	MoAb (anti-intraspecies CRI, paratope specific)	Ag-positive Ab$_3$ response genetically restricted	8, 56
FeLV	MoAb (VN)	MoAb (antiprivate, paratope specific)	Silent Id Ag-positive Ab$_3$ response	8, 57
Herpes simplex II	MoAb	Rabbit (paratope specific)	Decreased survival DTH: CTL	49
Reo III	MoAb	MoAb (paratope specific)	Ag-specific Ab$_3$ response DTH: CTL	40, 41
Sendai	— (T cell clone)	MoAb	Ag-specific Ab$_3$ response protection	42, 43, 45

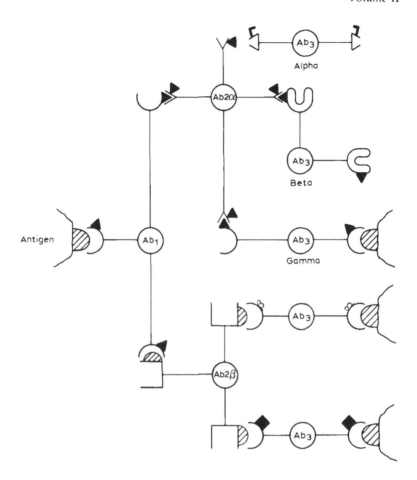

FIGURE 1.   Sets of antibodies involved in Id networks (Jerne 1974). (From UytdeHaag, F. G. C. M., Bunschoten, H., Weijer, K., and Osterhaus, A. D. M. E., *Immunol. Rev.*, 90, 93, 1986. With permission.)

rabbit Id.[9] Thus, this Ab_2, which is obviously an $Ab_{2\alpha}$, could overcome the genetic restrictions by activating "silent" interspecies cross-reactive idiotopes which are normally subject to down regulation. The coupling to appropriate carrier (LPS) may have been of critical importance.

The first demonstration of in vivo protection induced by the administration of Ab_2 came from a parasite system. Sacks et al.[10-12] showed that immunization with allogeneic Ab_2 antisera induced a genetically restricted protection against *Trypanosoma rhodesiense* infection. This protection was only found in mice bearing the Igh-c^a allotype and, thus, it may be concluded that this Ab_2 probably was not an internal image type of immunogen. In several virus models, the potential of $Ab_{2\alpha}$ and $Ab_{2\beta}$ for the modulation of specific immune response, resulting in the induction of, or the priming for, protective immunity, has been documented (Table 1).

The first report on a candidate internal image in a viral system came, also, from Urbain's group using TMV capsid protein as an antigen.[13] They described a rabbit Ab_2 which recognized an interspecies CRI present on anti-TMV antibodies from rabbits, mice, horses, goats, and chickens immunized with TMV. This Ab_2 induced an anti-TMV antibody response in mice, without the necessity of subsequent antigen boosting.

In the following section different approaches of using anti-Id antibody for the modulation of the immune response to a number of relevant human and animal virus infections are reviewed.

III. INDUCTION OF ANTIVIRAL B-CELL-MEDIATED IMMUNITY WITH Ab$_2$

A. Hepatitis B Virus

Hepatitis B virus (HBV), a member of the family Hepadnaviridae, is an important human pathogen which infects about 175 million people every year. Although most infected people recover completely from the acute hepatic disease, 5 to 10% develop a persistent carrier state, which may progress to hepatocellular injury, cirrhosis, and hepatocellular carcinoma. The majority of the estimated 200 million HBV carriers live in underdeveloped areas, which infection with the virus occurs at an early age. Antibodies directed to the group-specific determinant *a* of HBV surface antigen (HBsAg) are protective against infection, which is not the case with antibodies directed against the core antigen (HBcAg). Apart from the group-specific determinant, two sets of subtype determinants (*d, y, w,* or *r*) are recognized serologically, resulting in four major subtypes: *adw, ayw, adr,* and *ayr.* The development of a classical vaccine against HBV infection has long been impossible, since in vitro virus propagation systems were not available. Recently, vaccines, based upon purified 22-nm spherical HBsAg particles obtained from plasma of chronically infected individuals, have been shown to induce HBV neutralizing antibody (anti-HBs). Although the safety and efficacy of these vaccines in man have been documented in groups at risk of acquiring HBV infection,[14] the high cost and limited availability will restrict their potential as vaccines for the world-wide combat against HBV infection. Alternative approaches pursued for the generation of HBV vaccines based upon HBsAg utilize recombinant DNA or synthetic peptide technologies. Recently, different groups also initiated studies using Ab$_2$ in the HBV system, with the final goal to develop an internal image vaccine against HBV infection.

In studies concerning the Id networks in HBV infection, Kennedy and Dreesman[15] characterized a cross-reactive human anti-HBs Id by raising rabbit Ab$_2$ antisera. They detected this CRI in sera of most individuals who were vaccinated against, or naturally infected with HBV and showed, in inhibition assays with antigen, that it was located in or near the combining site of the antibody. This was further evidenced by showing that a cyclic synthetic peptide with a molecular weight of 1704 (peptide 1, analogous to the major polypeptide of HBsAg) could also inhibit the Id-anti-Id reaction. This cross-reactive anti-HBs Id proved to be directed against the conformation-dependent group-specific *a* determinant which is known to induce protective immunity in man and is present in all strains of HBV. An interspecies Id cross reaction intrinsically associated with anti-HBs-positive molecules could also be detected on anti-HBs sera from various mammalian species,[16] but not in similar antisera from chickens. Monoclonal anti-HBs antibodies that recognized determinants associated with cyclic peptide 1 all expressed the interspecies CRI. In vivo administration in mice or rabbit Ab$_2$ against the interspecies CRI resulted in an increased number of anti-HBs IgM-secreting spleen cells.[17] When the rabbit Ab$_2$ was not given in saline, but alum precipitated, also, an IgG response was observed. A single injection of the rabbit Ab$_2$ did not induce a detectable anti-HBs response and could only prime for subsequent boosting with HBsAg or peptide 1.

Immunization experiments with the Ab$_2$ in both rabbits and chimpanzees were conducted.[18] The Ab$_3$ response induced in one rabbit recognized the *a* determinant of HBsAg at a very low level. This level would probably be not protective against HBV infection. Two chimpanzees that received 1-mg injections of alum-precipitated rabbit Ab$_2$ produced detectable levels of anti-HBs. On the basis of the induction of anti-HBs antibody in another species with this rabbit Ab$_2$, reactive with anti-HBs antibodies from different species, they concluded that the Ab$_2$ represented a true internal image, although one would also have expected a reactivity with chicken anti-HBs sera. More generally, they concluded that the induction of an antibody response in a species susceptible to infection with human HBV indicates the potential of internal image-bearing vaccines for use against human disease.

Monoclonal Ab$_2$ mimicking HBsAg has also been described by Thanavalla et al.[19] Four mouse monoclonal Ab$_2$ preparations (MoAb$_2$) against the paratope of a mouse monoclonal anti-HBsAg antibody were generated. Two of these MoAb$_2$ recognized an interspecies CRI as evidence by the reactivities of these hybridomas in an IFA technique with sera from different mammalian species (goat, rabbit, swine, and human) immunized with HBsAg. No reactivity was observed with control and preinoculation sera lacking anti-HBsAg antibodies. Also, these MoAb$_2$ against interspecies CRI reacted with an idiotope recognizing the *a* determinant, since after selective absorption of the anti-*a* antibodies from swine antisera, their reactivity disappeared.

B. Poliovirus

The introduction of extensive vaccination programs in industrialized countries with either live-attenuated (Sabin) or inactivated (Salk) polio vaccines has had a great impact on the incidence of poliomyelitis, a human disease caused by poliovirus types I, II, or III. The disease almost completely disappeared from these countries, although especially in small groups of unimmunized individuals, endemic outbreaks have continued to occur. In the developing world, poliomyelitis is still a major problem, and an effective vaccine that can be produced at low cost for world-wide distribution, perhaps even with the final purpose to eradicate the disease, is urgently needed. A vaccine not containing live virus would be the best candidate to achieve this goal. Apart from developments in the field of candidate recombinant DNA and synthetic polio vaccines, which have not been too promising so far, the large-scale production of polio vaccines in a continuous monkey kidney cell line seems to be the best choice at this moment. However, the evaluation of the safety of these vaccines, especially with regard to the tumorogeneous potential of the production cells and the requirements of extensive purification methods to ensure the absence of nucleic acids, has not yet been fully made.

In an attempt to evaluate the feasibility of a MoAb$_2$ vaccine for poliomyelitis, we have chosen poliovirus type II as a model,[8,20,21] mainly because we had generated virus-neutralizing MoAb$_1$ which could protect mice after one single inoculation against a lethal intracerebral challenge with poliovirus type II. Moreover, there are only a limited number of neutralization-inducing epitopes on polioviruses[22,23] and humoral immunity probably constitutes the major component of protective immunity against infection with polioviruses.

From a panel of neutralizing BALB/c MoAb$_1$ generated against poliovirus type II (strain MEF$_1$), we selected three MoAb$_1$ (1-10C9, 4-15D4, and 11E7) on the basis of their broad reactivities in virus neutralization test with type II strains, but not with types I or III from all over the world. From these data and mutual inhibition studies with HRPO-labeled and -nonlabeled MoAb$_1$, it was concluded that these MoAb$_1$ define the same major neutralization inducing epitope on poliovirus type II. BALB/c mice were immunized with protein A-Sepharose®-purified MoAb$_1$, 1-10C9, and a panel of MoAb$_2$ was generated using an idiotope cross-linking system with plate-bound F(ab')$_2$ of MoAb$_1$ as a coat and a HRPO conjugate of MoAb$_1$ to detect MoAb$_2$ activity. One MoAb$_2$ (2-17C3) was selected, and the idiotope which it recognized on MoAb$_1$, 1-10C9 was found to be closely associated with its paratope, since the MoAb$_1$-MoAb$_2$ interaction could be inhibited by poliovirus type II (and not by type I or III). Preincubation of MoAb$_2$, 2-17C3 with either of the three selected MoAb$_1$, which were derived from two different fusions, caused a dose-dependent inhibition in the idiotope cross-linking ELISA. This could not be achieved with nonneutralizing antipoliovirus type II or antipoliovirus type I or III MoAb$_1$. Moreover, sera from BALB/c mice hyperimmunized with poliovirus type II also inhibited the idiotope cross-linking, while their preimmune sera failed to inhibit this reaction. Poliovirus type II-neutralizing hyperimmune sera of different animal species, including rats, guinea pigs, monkeys, and humans, but not their respective nonneutralizing preimmune sera, inhibited MoAb$_1$, 1-10C9 idiotope cross-

Table 2
MoAb$_2$, 2-17C3SCC INDUCE POLIOVIRUS-NEUTRALIZING ANTIBODIES IN BALB/c MICE

Immunogen	No. of mice	Poliovirus-neutralizing Ab	Survival after challenge (days)
MoAb$_2$, 2-17C3scc			
5 μg	2	32	5, 10
500 ng	2	32	10, 11
50 ng	2	<2	9, 12
5 ng	2	<2	8, 7
MoAb$_1$, 1-10C9E8[a]			
5 μg	2	>4096	>25
500 ng	2	2048	>25
50 ng	2	128	>25
5 ng	2	8, 2	15, 13

[a]   MoAb$_1$, 1-10C9E8 was used with a neutralizing titer of $10^{4.9}$/ 50 μℓ using 100 TCID$_{50}$. Mice were injected 24 hr prior to challenge with LD$_{50}$ of poliovirus type II strain MEF$_1$.

From UytdeHaag, F. G. C. M., Bunschoten, H., Weijer, K., and Osterhaus, A. D. M. E., *Immunol. Rev.*, 90, 93, 1986. With permission.

linking by MoAb$_2$, 2-17. These results showed that defined on an operational basis, the idiotope on MoAb$_1$, 1-10C9 is a cross-reactive idiotope associated with the virus-neutralizing antibody response to poliovirus type II in different mammalian species.

Immunization experiments with MoAb$_2$, 2-17C3 aiming at the induction of a protective immune response against poliovirus type II were carried out in BALB/c mice. They were immunized i.p. with various dilutions of purified MoAb$_1$, 1-10C9, MoAb$_2$, 2-17C3, or normal BALB/c IgG (Table 2). Animals inoculated with MoAb$_1$, 1-10C9 were challenged intracerebrally 24 hr later with 20 LD$_{50}$ poliovirus type II (strain MEF$_1$). Mice exhibiting virus neutralization antibody titers ≥128 survived this challenge, whereas control mice and mice exhibiting virus-neutralizing antibody titers <128 died within 15 days after challenge. The mice which had been inoculated with MoAb$_2$, 2-17C3 developed a dose-dependent Ab$_3$ response with 1-10C9 Id specificity. Virus-neutralizing antibody titers of 32 were detected in mice inoculated with 5 and 0.5 μg MoAb$_2$ in saline. However, these antibody levels did not protect them against intracerebral challenge with 20 LD$_{50}$ poliovirus type II (strain MEF$_1$). From these experiments, we concluded that MoAb$_2$, 2-17C3 acted without antigen by expanding MoAb$_1$, 1-10C9 interspecies cross-reactive idiotope bearing B-cell clones recognizing a major virus neutralization-inducing epitope of poliovirus type II.

C. Rabies Virus

The prevention and control of rabies mainly depends on epidemiologic measures like the impoundment of stray dogs and the immunization of domestic carnivores. Active immunization is one of the main measures taken to prevent rabies in man after exposure to the virus. Vaccines based upon inactivated whole virus produced in human diploid cells or primary animal cells are mainly used for this purpose at present, whereas similar vaccines produced in monkey kidney cell lines are being considered. The spike glycoprotein (G protein) of the virus has been shown to be involved in the generation of virus-neutralizing antibody, the generation of cell-mediated immunity,[24-26] and the induction of protective immunity. The G protein has been used to produce subunit vaccines (for review see Reference 27), and recently we have shown that a candidate immunostimulating complex (iscom) subunit rabies vaccine containing the G protein is capable of inducing virus-neutralizing antibody in vitro and in vivo, cell-mediated immunity, and protection of mice against intracerebral challenge with rabies virus.[28] The G protein has been expressed in a vaccinia virus vector system and it has been shown that this candidate rabies vaccine also induces virus-neutralizing antibody, stimulates cytotoxic T cells, and protects against intracerebral challenge.[29]

The potential of Ab_2 to provide protective immunity against rabies virus was first studied by Reagan et al.[30] With the aim of producing F polyclonal Ab_2, rabbits were immunized with protein A-Sepharose® chromatography purified mouse $MoAb_1$ against nonoverlapping epitopes of the G protein, which recognized major neutralization-inducing sites on rabies virus. The absence of significant cross reactivity among the rabbit Ab_2 for heterologous $MoAb_1$ suggested that they reacted highly specific with unique variable region determinants. The binding of three of the five rabbit Ab_2 to their homologous $MoAb_1$ could be inhibited by soluble G protein, indicating a reactivity with paratope determinants. Immunization of ICR randomly bred mice with protein A-Sepharose® purified rabbit Ab_2 preparations resulted with two of the five preparations in the induction of a specific virus-neutralizing antibody response. One of these rabbit Ab_2 elicited an Ab_3 response directed against the same major antigenic site involved in virus neutralization as detected by the corresponding $MoAb_1$. Since the Id marker expressed on this $MoAb_1$ is not expressed in human or rabbit rabies virus immune sera, but is present in the sera of BALB/c mice, the authors concluded that this idiotope is not interspecies cross reactive, which makes the rabbit Ab_2 against this idiotope an unlikely candidate for an internal image. The virus-neutralizing antibody titer in the Ab_2-immunized mice proved not high enough to protect against lethal intracerebral challenge with rabies virus. In order to determine the significance of these low virus-neutralizing antibody titers, mice were immunized with a mixture of the three rabbit Ab_2, 3 weeks prior to rabies vaccination. By showing a clear booster effect in the development of virus-neutralizing antibody titers, it was concluded that the immune system of these mice had been sensitized to subsequent antigen exposure. This effect was confirmed by showing the induction of a dose-dependent Ab_2 manipulation of specific immunity resulting in an increased survival in an intracerebral challenge experiment in mice.

We have prepared $MoAb_2$ against mouse $MoAb_1$ antirabies virus G protein.[8] For this purpose we used two virus-neutralizing $MoAb_1$ (1-10B8 and 1-11D6) which we had raised to antigenic site 1 of the G protein, as was shown in competition assays using the same panel of $MoAb_1$ used by Reagan et al. which were generously provided by Dr. T. Wiktor, Wistar Institute, Philadelphia. Protein A-Sepharose® chromatography-purified $MoAb_1$, 1-10B8 was coupled to KLH and used for immunization of BALB/c mice to generate $MoAb_2$. Hybridoma supernatants were screened on a $F(ab')_2$ coat of $MoAb_1$, 1-10B8 using an antimouse Fc-HRPO conjugate. One $MoAb_2$, 2-23 was selected for further studies. We showed that the homologous $MoAb_1$, 1-10B8 and the heterologous $MoAb_1$, 509-6 obtained from Dr. T. Wiktor could completely inhibit the binding of $MoAb_2$, 2-23 to $MoAb_1$, 1-10B8 or $MoAb_1$, 1-11D6. Other $MoAb_1$ directed against other antigenic sites of rabies virus did not inhibit this reaction. Moreover, hyperimmune sera from outbred NIH mice and human individuals immunized with rabies vaccine did not express the 1-10B8 idiotope, although antiantigenic site I activity was present. Since the three $MoAb_1$ used were derived from different fusion events, it was concluded that the $MoAb_1$, 1-10B8 idiotope is expressed in BALB/c mice in a recurrent fashion and in association with a B-cell response against antigenic site I of rabies virus G protein. The possibility of a genetic restriction of the expression of

Table 3

**INDUCTION OF ANTIRABIES VIRUS ANTIBODIES IN
BALB/c MICE BY MoAb$_2$, 2-23: EFFECT OF DOSAGE AND
MANNER OF PRESENTATION**

Dosage of MoAb$_2$, 2-23 administered (μg)	Antibody response (OD450 nm $\times$ 10^3)			
	Ab$_3$ response to MoAb$_2$,[a] 2-23 given in		Antirabies virus[b] response to MoAb$_2$, 2-23 given in	
	Saline[c]	KLH[d]	Saline	KLH
200	≤250	1251	≤30	89
	≤250	1397	≤30	346
	≤250	1599	≤30	149
20	≤250	1217	≤30	104
	267	1228	≤30	81
	1065	N.D.	376	323
2	717	≤250	143	≤30
	≤250	≤250	49	≤30
	≤250	292	≤30	≤30

[a]   ELISA: binding of mouse serum antibody to plate-bound MoAb$_2$, 2-23 F(Ab)$_2$ detected by goat antimouse IgG (Fc fragment specific)-HRPO.

[b]   ELISA: binding of mouse serum antibodies to plate-bound rabies virus detected by goat antimouse IgG (Fc fragment specific)-HRPO.

[c]   Two injections of 100, 10, or 1 μg were given i.p.

[d]   Two injections of 100, 10, or 1 μg were given s.c. in CFA and IFA, respectively.

From UytdeHaag, F. G. C. M., Bunschoten, H., Weijer, K., and Osterhaus, A. D. M. E., *Immunol. Rev.*, 90, 93, 1986. With permission.

the MoAb$_1$, 1-10B8 idiotope by MHC, Igh-C, or Igh-V genes, which has also been reported for the expression of recurrent idiotopes in other systems (for review see Reference 8), was studied by analysis of the immune responses of different inbred and congeneic strains of mice to rabies virus. All the mice developed antisite 1 antibody in response to rabies virus. MoAb$_1$, 1-10B8 idiotope-positive clones were only observed in BALB/c (H-2^d, IghC^{a}V^a), BALB/B (H-2^b, IghC^{a}V^a), and BALB.14 (H-2^d, IghC^{b}V^a) mice, but not in C57BL/6 (H-2^d, IghC^{b}V^b), C.B20 (H-2^d, IghC^{b}V^b), or BALB/Ig^b (H-2^d, IghC^{b}V^b) mice. From these results it was concluded that the activation of MoAb$_1$, 1-10B8 idiotope-bearing clones by rabies virus is associated with the IghV gene complex. Immunization studies were carried out in BALB/c mice with doses of MoAb$_2$, 2-23 ranging from 2 to 200 μg, either in saline or coupled to KLH (Table 3). A dose-dependent Ab$_3$ (anti-MoAb$_2$, 2-23) response was found in mice inoculated with 2 or 20 μg MoAb$_2$, 2-23 in saline. Injection with MoAb$_2$ coupled to KLH resulted in the induction of a strong Ab$_3$ response, which was accompanied by antirabies virus binding antibody. Immunization of C57BL/6 and CB20 mice with 100 μg of MoAb$_2$, 2-23-KLH did not induce antirabies virus antibodies, which supported the IghV^a gene linkage of the expression of the MoAb$_1$, 1-10B8 idiotope expression in a rabies virus-induced anti-G site 1 antibody response.

D. Feline Leukemia Virus: A Retrovirus Model

Leukemia is the most common neoplastic disease in cats and is caused by a horizontally transmitted C-type retrovirus, feline leukemia virus (FeLV).[32] Cats develop either a transient or a persistent infection after natural infection. Persistently infected animals are prone to degenerative, proliferative, or neoplastic changes of cells of the hematopoietic series, resulting in immunosuppression, leukemia, or, less frequently, fibrosarcoma.[33,34] Transiently infected cats may develop protective immunity. The pathogenesis and clinical syndrome of the immunosuppressive disease shows remarkable similarities with acquired immunodeficiency syndrome (AIDS) of man, which is also caused by a member of the retroviridae family: HIV(LAV/HTLVIII). The urgent need for the development of a vaccine against AIDS in man has prompted us to use the FeLV-cat system as a model to explore different approaches leading to the immunoprophylaxis of cats against FeLV-associated disease. It has been shown that polyclonal antibodies against the virus envelope glycoprotein — gp70/ 85 — which neutralize the virus may protect cats against FeLV infection, even when administered shortly after exposure.[35] Postexposure administration of virus-neutralizing mouse monoclonal antibody directed against one epitope of gp70 failed to protect cats against FeLV infection.[36] Several methods have been considered for the vaccination of cats against FeLV infection, including live or inactivated whole virus or virus-producing cells. A FeLV subunit vaccine consisting of the envelope glycoprotein of the virus is ideally the safest (i.e., genome-free) type of vaccine. Most of the experimental work on subunit vaccines against mammalian retroviruses has been carried out with gp71/85 of Friend murine leukemia virus (F.MuLV). The initial limited success in developing these subunit vaccines probably rested in the use of the protein in its monomeric form. Studies with F-MuLV showed that micellar preparation to the glycoprotein could prevent erythroleukemia in mice.[37] We recently described a novel structure for the antigenic presentation of membrane proteins from enveloped viruses: the immunostimulating complex (iscom).[38] These iscom preparations proved highly immunogenic and induced an antibody response even to membrane proteins that are considered poorly immunogenic. We described the preparation of an iscom from the gp70/85 of FeLV and the induction of a protective immune response in cats with this candidate FeLV subunit vaccine.[39]

Among the approaches to develop a FeLV vaccine, we also considered the generation of an anti-Id vaccine. A panel of mouse MoAb$_1$ was prepared against gp70/85, using spleen cells from BALB/c mice immunized with gradient-purified FeLV. Hybridomas were selected which produce virus-neutralizing anti-gp70≡85 MoAb$_1$ reactive with the three subtypes FeLV-A, FeLB-B, and FeLV-C, as shown in ELISA, virus neutralization, immunofluorescence, and immunoblotting assays. Three MoAb$_1$ derived from one fusion were selected: MoAb$_1$, 3-17; MoAb$_1$, 6-15; and MoAb$_1$, 7-3. In competition assays with horseradish peroxidase (HRPO)-labeled MoAb$_1$, it was shown that these MoAb$_1$ all recognized the same or spatially related virus neutralization-inducing site on gp70≡85 (Table 4). For the production of a panel of MoAb$_2$, BALB/c mice were immunized with MoAb$_1$, purified by protein A-Sepharose® chromatography, and coupled to KLH. The fusions of spleen cells of four mice were screened on a MoAb$_1$ (F(ab')$_2$ coat using a HRPO-labeled antimouse Fc preparation to show the presence of MoAb$_2$. Fifteen stable MoAb$_2$-producing hybridomas were obtained. The binding of all these MoAb$_2$ to MoAb$_1$, 3-17 could be inhibited by gradient-purified FeLV, suggesting that they recognized an idiotope located within the paratope of MoAb$_1$, 3-17. The binding of all 15 MoAb$_2$ to plate-bound MoAb$_1$, 3-17 could be inhibited by preincubation of MoAb$_2$ with either of the MoAb$_1$, 3-17, 6-15, or 7-3, but not by preincubation with FeLV hyperimmune virus-neutralizing sera of BALB/c mice, rats, guinea pigs, cats, goats, or rabbits. Since all three MoAb$_1$ used resulted from the same fusion event, we concluded that all MoAb$_2$ were directed against private idiotopes of MoAb$_1$. In competition studies in which the reactivity of each of the HRPO-labeled MoAb$_2$, with plate-bound MoAb$_1$, 3-17, was inhibited with each of the nonlabeled MoAb$_2$, it was shown that these MoAb$_2$ recognized two different partially overlapping idiotopes in the paratope of MoAb$_1$, 3-17.

Since Ab$_{2\alpha}$ with specificity for either intrastrain recurrent or private idiotopes have been

Table 4
CHARACTERISTICS OF ANTI-FeLV MoAb₁

MoAb	ELISA[a]	Neutralization[b]			Fluorescence[c]			Specificity[d]
		A	B	C	A	B	C	
7-3[e]	1.2×10^5	5120	2560	10^4	≥5	≥5	≥5	Anti-GP70
6-15[e]	2.5×10^5	1280	2560	2560	≥5	≥5	≥5	Anti-GP70
3-17[e]	5×10^5	5120	5120	5120	≥5	≥5	≥5	Anti-GP70

[a] ELISA on microtiter plates coated with 0.1 μg FeLV/well.
[b] Neutralization focus assay with MSV (FeLV-A, -B, or -C) as virus.
[c] Membrane fluorescence on FL74 cells (containing FeLV subgroups A, B, and C) as targets. Titer = log 10.
[d] Determined by Western blot analysis with purified FeLV analyzed in 7.5% SDS-PAGE, 20 μℓ of MoAb, and goat antimouse IgG (Fc fragment specific) peroxidase conjugate.
[e] MoAbs derived from the same fusion.

From UytdeHaag, F. G. C. M., Bunschoten, H., Weijer, K., and Osterhaus, A. D. M. E., *Immunol. Rev.*, 90, 93, 1986. With permission.

shown to be capable of eliciting the synthesis of Ab₃ responses with predefined Ab₁-like antigen specificity, even across allotype linkage or species barriers (for review see Reference 8) we have attempted to induce the generation in BALB/c mice of Ab₃ sharing MoAb₁, 3-17 anti-FeLV activity using a selection of these MoAb₂ in different immunization protocols. In a first experiment (Table 5) 10 μg of MoAb₂, 2-17 in saline, copolymerized with LPS or coupled to KLH, was inoculated i.p. at weekly intervals. Eight weeks after the immunization had started only the mice inoculated with the LPS or KLH-coupled MoAb₂ had developed an Ab₃ response (Table 4). Two mice inoculated with MoAb₂, 2-17-LPS were boosted with the same material 20 weeks later. One of them developed a low level of anti-FeLV antibody in an ELISA. This indicates that, as has been reported by Francotte and Urbain,[9] silent idiotopes of predefined antigen specificity in absence of antigen may be induced with Ab₂ defining a private paratope-related idiotope on MoAb₁ recognizing gp70/85 of FeLV. At present we are investigating whether sequential immunization with the different MoAb₂ defining nonoverlapping idiotopes of MoAb₁, 3-17 will expand more selectively B-cell precursors of the desired anti-FeLV specificity. Furthermore, we are investigating whether the administration of FeLV to MoAb₂, 2-17-treated mice will induce the appearance of FeLV-binding antibody sharing idiotopes with MoAb₁, 3-17.

IV. INDUCTION OF ANTIVIRAL T-CELL-MEDIATED IMMUNITY WITH Ab₂

All systems described so far have been focused on the generation of Ab₃ responses sharing idiotope expression and antigen binding capacity with the initial Ab₁-producing B cells. Recently, evidence has been acquired in two animal virus systems indicating that B- and T-cell-mediated protective immunity may be induced against viruses by administration of MoAb₂ or anticlonotypic antibody against variable region determinants of the T-cell receptor of cloned T cells.

A. Reovirus Type III

Sharpe et al.[40] generated a MoAb₂ in BALB/c mice against a BALB/c MoAb₁ directed to the hemagglutinin of reovirus type III and exhibiting virus-neutralizing activity. This MoAb₂ was shown to block virus binding to neuronal cells and CTL, cytotoxic T-cell-

Table 5
IMMUNIZATION OF BALB/c MICE WITH
MoAb$_2$, 2-17

	Ab$_3$response (OD 450 nm × 10^3)		
	anti-MoAb$_2$, 2-17		
Immunogen	Pre	Post (n = 5)	Anti-FeLV
40 μg MoAb$_2$ LPS[a]	≤40	>1400	None
40 μg MoAb$_2$, KLH	≤40	>1400	None
40 μg MoAb$_2$	≤40	≤40	None
LPS/KLH	≤40	≤40	None

Note: Mice were injected i.p. four times at weekly intervals with 10 μg MoAb$_2$, 2-17 either in saline or copolymerized with LPS and with 10 μg LPS alone.

Sera were analyzed before immunization and 2 weeks after the last immunization for expression of anti-MoAb$_2$, 2-17 activity in ELISA and for anti-FeLV antibodies in ELISA's and membrane immunofluorescence. No antibody activity against an irrelevant MoAb of the same subclass (IgGl, κ) was detectable (dilution of sera 1:45).

[a] Boosted 12 weeks later with 10 μg MoAb$_2$, LPS → 30 (anti-FeLV).

From UytdeHaag, F. G. C. M., Bunschoten, H., Weijer, K., and Osterhaus, A. D. M. E., *Immunol. Rev.*, 90, 93, 1986. With permission.

mediated lysis of virus-infected target cells, and to induce virus-specific VN antibody in mice.[41] S.c. immunization of mice with 100 μg of this MoAb$_2$ elicited a dose-dependent delayed-type hypersensitivity (DTH) response after virus challenge in the footpad. This MoAb$_2$-elicited DTH could be adoptively transferred with T cells. It proved specific for the neutralization-inducing domain of reovirus type 3 hemagglutinin, since it could not be induced by reovirus type 3 variants lacking the relevant HA epitope. The extent of the DTH responsiveness to MoAb$_1$ was not restricted by the IgH locus.

Furthermore, T cells from the spleens of mice, immunized s.c. with irradiated Ab$_2$-bearing hybridoma cells, generated vigorous cytotoxic activity against reovirus-infected target cells. These studies clearly demonstrated that nonadjuvated purified MoAb$_2$, induced by MoAb$_1$ directed to the reovirus HA-neutralizing domain, can induce potent reovirus-specific T-cell immunity in the absence of antigen. The induction of a CTL-mediated response with soluble MoAb$_2$ was much less efficient, which may indicate the requirement of MHC signals for activation of CTL.

Thus, MoAb$_2$ generated against a B-cell idiotope may induce a T-cell response, which suggests a sharing of the idiotypic repertoire between B- and T-cell receptors.

B. Sendai Virus

Another approach was followed by Ertl and Finberg who used a Sendai virus-specific T-helper cell clone[42] to generate an anticlonotypic antibody.[43] This clone was shown to be Lyt-1.2$^-$, Lyt-2.2$^-$, and Thy-1.2$^+$, proliferated and secreted lymphokines in response to Sendai virus presented on IA^d-positive stimulator cells, and mediated a virus-specific DTH. Although injection of this clone was shown to increase in vivo a cytolytic T-cell response to Sendai virus, it failed to protect mice against lethal challenge with the virus. The syngeneic anticlonotypic monoclonal antibody recognized 30% of Sendai virus-specific T-cell clones or lines, and the expression of this clonotype proved genetically not restricted, since it was found on T cells of mouse strains regardless of their H-2 haplotype or Igh allotype. It was concluded that the antibody recognizes a dominant idiotope of the Sendai virus-specific T-cell repertoire. In vitro, this antibody stimulated Sendai virus-specific cytolytic T cells. In vivo the anticlonotypic antibody induced T cells which mediated a DTH, with a remarkable lack of H-2 restriction as compared to Sendai virus-induced T cells, suggesting that the antibody stimulates T cells with a distinct specificity which is not triggered upon immunization with the viral antigen. It also induced effector cells which, upon restimulation in vitro with Sendai virus, lysed Sendai virus-infected target cells as well as the anticlonotypic-expressing hybridoma cell line. Also, the anticlonotypic antibody-induced cytolytic T cells were less H-2 restricted as compared to antigen-induced T cells. Antigen and anticlonotypic antibody induced T cells and lysed the hybridoma cell line independent of corecognition of H-2; allogeneic effector cells proved nearly as effective as H-2-compatible effector cells. Consequently, the affinity of the T-cell receptor to the anticlonotypic antibody expressed as surface Ig on the hybridoma was sufficient to trigger lysis.

Moreover, the anticlonotypic antibody directed against a Sendai virus-specific T-helper cell clone also induced a Sendai virus-specific antibody response as tested by ELISA, in inbred mice of different H-2 allotype or Igh allotype. The response seemed under T-cell control, since nude mice failed to respond to stimulation with the anticlonotypic antibody or Sendai virus. Finally, it was shown that mice immunized with purified anticlonotypic antibody were protected against lethal challenge with Sendai virus.

V. SUMMARY AND DISCUSSION

Based upon the proposal of Jerne's network hypothesis of the immune system,[3] the concept of using idiotope structures for vaccination purposes, initially suggested by Nisonoff and Lamoyi,[6] has been considered for the generation of vaccines against infectious agents, but also for the manipulation of the immune response in neoplastic and autoimmune diseases and for the modulation of graft-vs.-host reactions (for review see Reference 44). The potential of the exploitation of idiotope structures for the modulation of the immune response to infectious agents was first indicated by Sacks et al.[10,12] who induced in vivo protective immunity in mice against *Trypanosoma rhodesiense* infection with Ab_2 not displaying internal image properties. A growing reluctance to use live virus vaccines is based upon a declined acceptability of adverse side effects in a situation of decreased incidence of the infection and the fear of inducing latency or malignancy with members of certain virus families. These considerations, the lack of suitable production systems for certain viruses, and the failure of most nonlive virus vaccines to induce long-lasting and T-cell-mediated immune responses have prompted the search for novel approaches. In several virus systems the use of anti-Id antibodies for vaccination purposes has now been studied. Although, in most cases, the immune responses elicited by Ab_2 were inferior to those obtained with the whole conventionally presented virus or viral antigen, one should realize that, so far in most systems, only Ab_2 preparations have been used which are specific for one idiotypic determinant on an Ab_1, with specificity for only one antigenic determinant of the virus. In several virus systems Ab_2 in the absence of antigen was able to elicit virus-neutralizing antibody responses and T-cell-mediated responses (CTL and DTH) which seemed less MHC restricted than those generated by antigens presented in the conventional way. The immune responses generated were, in most cases, too low to be protective, although in certain systems an effective priming effect or even a direct protection was observed. The use of Ab_2 preparations may also prove valuable in situations of immune incompetence, when a given individual is not able to respond adequately to a given pathogen.[45] This was, e.g., shown by Stein and Söderstrom when they demonstrated that neonates which cannot generate an effective immune response to *Escherichia coli* may be primed for a subsequent contact with *E. coli* by inoculation with an Ab_2 directed to an antibody to *E. coli* capsular polysaccharide.[46] Also, the activation of silent Id of predefined antigen specificity across a species barrier using polyclonal xenogeneic Ab_2 defining a private Id, as was shown by Francotte and Urbain,[9] is indicative for this potential of Ab_2. On the other hand, it has been suggested that Ab_2 are not only capable of enhancing immune responses, but may under certain conditions also turn off immune responses.[14,47,48] So it was shown by Kennedy et al. that the mortality of mice who had received rabbit Ab_2 preparation generated against a $MoAb_1$ to herpes simplex virus type II (HSV-2) was increased in comparison with mice which had received an irrelevant antibody.[49] In the future one may be able to exploit the use of Ab_2 to prevent certain unwanted (auto-)immune reactions or turn off unwanted consequences of vaccines or virus infections.

In order to avoid problems with genetic restriction, obviously the best candidates for vaccination are true internal images. In several viral systems true or candidate internal images have been generated and tested for their immunogenic potential: the rabbit polyclonal Ab_2 used by Kennedy and Dreesman, and the $MoAb_2$ used by Thanavalla et al. in the hepatitis B system; a rabbit polyclonal Ab_2 to an anti-VEE virus $MoAb_1$ which induces virus-neutralizing antibody;[50] the $MoAb_2$, used by Sharpe et al. in the reovirus III system, the monoclonal anticlonotypic antibody used by Ertl and Finberg in the Sendai system, and the $MoAb_2$ which we used in the poliovirus type II system, may all represent internal images of external epitopes of the respective viruses. All these preparations were shown not only to elicit an Ab_3 (anti-Ab_2) response, but also to elicit an antibody response recognizing the virus. In certain cases, only a booster effect after subsequent antigen exposure was noted, or in other cases too low virus-neutralizing antibody concentrations were induced to provide effective protection against challenge with the virus. However, in the reovirus III system and in the Sendai virus system, it was shown that $MoAb_2$ directed to either a B- or a T-cell idiotope induced both a B- *and* T-cell-mediated antiviral immunity, which in the latter also proved to be protective against virus challenge. In virus systems where Ab_2 were used not displaying $Ab_{2\beta}$ properties, like the $MoAb_2$ which we used in the rabies virus system, a genetically restricted antivirus antibody response was observed. In the FeLV system we generated $MoAb_2$ against private idiotopes of a $MoAb_1$, which within a syngeneic BALB/c system after prolonged immunization protocols with LPS-copolymerized antibody preparations induced low-level anti-FeLV antibody. This suggests that in this way, silent idiotopes of predefined antigen specificity can be induced.

From the data presented in various virus systems, it has become clear that apart from immunization schedules used, also the "antigenic" presentation of the idiotypic structures may be of crucial importance. It was clearly shown in the HBV system by Kennedy and Dreesman[51] and by us, both in the rabies virus and the FeLV system, that multimeric presentation forms of Ab_2 were more efficient in eliciting an Ab_3 and/or an antiviral response. Furthermore, Sharpe et al. showed in the reovirus III system that $MoAb_2$-bearing hybridoma cells, in contrast to soluble $MoAb_2$, generated vigorous CTL-mediated activity. This may also relate to MHC Class I and II antigen expression on hybridoma cells required for activation of this set of cells,[40] which would be in agreement with the observation by Kaufmann et al.[52] who showed that a multimeric presentation of Id in combination with MHC Class I and II determinants was more efficient.

So far, we have only discussed the use of Ab_2 as potential vaccines. However, in various systems it has been suggested that the administration of Ab_1 resulted not only in an Ab_2 response, but also in an Ab_3 response, displaying Ab_1 specificity.[53] Interesting in this light is the suggestion of Reagan et al.[30] that the protective effect of a postexposure antirabies treatment with polyclonal human hyperimmune serum may be due to the generation of an antivirus Ab_3 response rather than to the direct antiviral effect of Ab_1. We have observed that immunization of rabbits with antiviral $MoAb_1$ either to rabies virus or to FeLV resulted in the generation of rabbit antibody directed against the virus.[56,57] Consequently, the idea of using Ab_1 for the induction of specific immune response against infectious agents should also be considered. Recently, Kang and Kohler described an antibody that mimics the antigen of its own paratope, i.e., an antibody with complementary paratope and idiotope.[54] They proposed the term "autobody" for this type of antibody which may have an important role in autoimmunity and a unique role in the network of the immune system. In fact, "autobodies" may prove ideal candidate molecules for Id vaccines. They may, on the one hand, transfer immediate passive immunity via the antibody quality, and active immunity by virtue of being an antigen image.

For practical reasons, Id vaccines should be prepared from monoclonal antibodies rather than from polyclonal antibodies. The production of polyclonal antibodies is difficult to standardize and only a minority of these antibodies prepared in animals would trigger Id-bearing cells with specificity for the antigen (for review see Reference 10). Moreover, developments in the hybridoma technology (for review see Reference 55) will permit the large-scale production of MoAb from different species (including man) of desired specificities. Id vaccines should contain panels of MoAb, corresponding with different epitopes on the virus. After the determination of relevant idiotypic structures for vaccination on MoAb, these structures may eventually be combined, perhaps even within one molecule, by recombinant DNA technology, or after sequence data have been obtained be constructed as synthetic peptides.

ACKNOWLEDGMENTS

We are greatly indebted to Jacques Urbain and co-workers for advice and stimulating discussions. We thank our students, Ch. Bodar, M. Jansen, J. Jacobs, F. Heemskerk, J. Wagenaar, and T. Glaudemans, for their important contributions to the results presented in this paper. The expert technical assistance of G. Drost, R. Bakker, K. Siebelink, and H. Loggen is greatfully acknowledged. We wish to thank C. Kruyssen for preparing the manuscript. This work was supported in part by a fellowship of the World Health Organization and by a grant (8503) from the European Veterinary Laboratory, Amsterdam, The Netherlands.

REFERENCES

1. **Norrby, E.,** Viral vaccines: the use of currently available products and future developments, *Arch. Virol.,* 76, 163, 1983.
2. **Quinnan, G. V., Ed.,** Vaccinia virus as vectors for vaccine antigens, in *Proc. Workshop on Vaccinia Viruses as Vectors for Antigens,* Elsevier, New York, 1985.
3. **Jerne, N. K.,** Towards a network theory of the immune system, *Ann. Immunol. (Paris),* 125C, 373, 1974.
4. **Kohler, G. and Milstein, C.,** Continuous cultures of fused cells secreting antibody of predefined specificity, *Nature (London),* 256, 459, 1975.
5. **Lindenman, J.,** Homobodies: do they exist?, *Ann. Immunol. (Paris),* 130C, 311, 1978.
6. **Nisonoff, A. and Lamoyi, E.,** Implications of the presence of an internal image of the antigen in anti-idiotypic antibodies: possible application to vaccine production, *Clin. Immunol. Immunopathol.,* 21, 397, 1981.
7. **Sege, K. and Peterson, P. A.,** Use of anti-idiotypic antibodies as cell-surface receptor probes, *Proc. Natl. Acad. Sci. U.S.A.,* 75, 2443, 1978.
8. **UytdeHaag, F. G. C. M., Bunschoten, H., Weijer, K., and Osterhaus, A. D. M. E.,** From Jenner to Jerne: towards idiotype vaccines, *Immunol. Rev.,* 90, 93, 1986.
9. **Francotte, M. and Urbain, J.,** Induction of anti-tobacco mosaic virus antibodies in mice by rabbit anti-idiotypic antibodies, *J. Exp. Med.,* 160, 1485, 1984.
10. **Sacks, D. L., Kelsoe, G. H., and Sacks, D. H.,** Induction of immune responses with anti-idiotypic antibodies: implications for the induction of protective immunity, *Springer Semin. Immunopathol.,* 6, 79, 1983.

11. **Sacks, D. L.,** Molecular mimicry of parasite antigens using anti-idiotypic antibodies, *Curr. Top. Microbiol. Immunol.,* 119, 45, 1985.
12. **Sacks, D. L. and Sher, A.,** Evidence that anti-idiotype induced immunity of African trypanasomiasis is genetically restricted and requires recognition of combining site related idiotyopes, *J. Immunol.,* 131, 1511, 1983.
13. **Urbain, J., Wikler, M., Franssen, J. D., and Collignon, C.,** Idiotypic regulation of the immune system by the induction of antibodies against anti-idiotypic antibodies, *Proc. Natl. Acad. Sci. U.S.A.,* 74, 5126, 1977.
14. **Hilleman, M. R., Buynbak, E. B., McAleer, W. J., McLean, A. A., Provost, P. J., and Tytell, A. A.,** Hepatitis A and hepatitis B vaccines, in *Viral Hepatitis,* Szmuness, W., Alter, H. J., and Maynard, J. E., Eds., Franklin Institute Press, Philadelphia, 1982, 385.
15. **Kennedy, R. C. and Dreesman, G. R.,** Common idiotypic determinant associated with human antibodies to hepatitis B surface antigen, *J. Immunol.,* 130, 385, 1983.
16. **Kennedy, R. C., Ionescu-Matiu, I., Sanchez, Y., and Dreesman, G. R.,** Detection of interspecies idiotypic cross-reactions associated with antibodies to hepatitis B surface antigen, *Eur. J. Immunol.,* 13, 232, 1983.
17. **Kennedy, R. C., Adler-Storthz, K., Henkel, R. D., Sanchez, Y., Melnick, J. L., and Dreesman, G. R.,** Immune response to hepatitis B surface antigen: enhancement by prior injection of antibodies to the idiotype, *Science,* 221, 853, 1983.
18. **Kennedy, R. C., Eichberg, J. W., and Dreesman, G. R.,** Lack of genetic restriction by a potential anti-idiotype vaccine for type B viral hepatitis, *Virology,* 148, 369, 1986.
19. **Thanavalla, Y. M., Bond, A., Tedder, R., Hay, F. C., and Roitt, I. M.,** Monoclonal "internal image" antiidiotypic antibodies of hepatitis B surface antigen, *Immunology,* 55, 197, 1985.
20. **UytdeHaag, F. G. C. M. and Osterhaus, A. D. M. E.,** Induction of neutralizing antibody in mice against poliovirus type II with monoclonal anti-idiotypic antibody, *J. Immunol.,* 134, 1225, 1985.
21. **UytdeHaag, F. G. C. M. and Osterhaus, A. D. M. E.,** Vaccines from monoclonal anti-idiotypic antibody: poliovirus infection as a model, *Curr. Top. Microbiol. Immunol.,* 119, 31, 1985.
22. **Fergusson, M., Minor, P. D., Margrath, D. I., Qui Yi-Hua, Spitz, M., and Schild, G. C.,** Neutralizing epitopes on poliovirus type 3 particles: an analysis using monoclonal antibodies, *J. Gen. Virol.,* 65, 197, 1984.
23. **Evans, D. M., Minor, P. D., Schild, G. S., and Almond, J. W.,** Critical role of an eight amino acid sequence of VP1 in neutralization of poliovirus type III, *Nature,* 304, 459, 1983.
24. **Cox, J. H., Dietzschold, B., and Schneider, L. G.,** Rabies virus glycoprotein. II. Biological and serological characterization, *Infect. Immun.,* 16, 754, 1977.
25. **Wiktor, T. J., Kamo, I., and Koprowski, H.,** In vitro stimulation of rabbit lymphocytes after immunization with live and inactivated rabies vaccines, *J. Immunol.,* 112, 2013, 1974.
26. **Macfarlan, R. I., Dietzschold, B., Wiktor, T. J., Kiel, M., Houghten, R., Lerner, R. A., Sutcliffe, J. G., and Koprowski, H.,** T cell responses to cleaved rabies virus glycoprotein and to synthetic peptides, *J. Immunol.,* 133, 2478, 1984.
27. **Wunner, W. H., Dietzschold, B., Curtis, P. J., and Wiktor, T. J.,** Review article: rabies subunit vaccines, *J. Gen. Virol.,* 64, 1649, 1983.
28. **Osterhaus, A., Sundquist, B., Morein, B., van Steenis, G., and UytdeHaag, F.,** Comparison of an experimental rabies virus iscom subunit vaccine with inactivated dog kidney cell vaccine, Abstract, in 5th Int. Conf. on Comparative Virology, Alberta, Can., May 4 to 10, 1986.
29. **Wiktor, T. J., Macfarlan, R. I., Reagan, K. J., Dietzschold, B., Curtis, P. J., Wunner, W. H., Kieny, M. P., Lathe, R., Lecocq, J. P., Mackett, M., Moss, B., and Koprowski, H.,** Protection from rabies by a vaccinia virus recombinant containing the rabies virus glycoprotein gene, *Proc. Natl. Acad. Sci. U.S.A.,* 81, 7194, 1984.
30. **Reagan, K. J., Wunner, W. H., Wiktor, T. J., and Koprowski, H.,** Anti-idiotypic antibodies induce neutralizing antibodies to rabies virus glycoprotein, *J. Virol.,* 48, 660, 1983.
31. **Reagan, K. J.,** Modulation of immunity to rabies virus induced by anti-idiotypic antibodies, *Curr. Top. Microbiol. Immunol.,* 119, 15, 1985.
32. **Hardy, W. D., Jr., Old, L. J., Hess, P. W., Essex, M., and Cotter, S.,** Horizontal transmission of feline leukemia virus, *Nature,* 244, 266, 1973.
33. **Hardy, W. J., Jr., Geering, G., Old, L. J., de Harven, E., Brodey, R. S., and McDonough, S.,** Feline leukemia virus: occurrence of viral antigen in the tissues of cats with lymphosarcoma and other diseases, *Science,* 166, 1019, 1969.
34. **Pedersen, N. C., Theilen, G. H., Keane, M. A., Fairbanks, L., Mason, T., Orser, B., Chen, C. H., and Allison, C.,** Studies of naturally transmitted feline leukemia virus infection, *Am. J. Vet. Res.,* 38, 1523, 1977.

35. **Hoover, E. A., Schaller, J. P., Mathes, L. E., and Olsen, R. G.,** Passive immunity to feline leukemia. Evaluation of immunity from dams naturally infected and experimentally vaccinated, *Infect. Immun.,* 16, 54, 1977.

36. **Weijer, K., UytdeHaag, F. G. C. M., Jarrett, O., Lutz, H., and Osterhaus, A. D. M. E.,** Post-exposure treatment with monoclonal antibodies in a retrovirus system: failure to protect cats against feline leukemia virus infection with virus neutralizing monoclonal antibodies, *Int. J. Cancer,* 38, 81, 1986.

37. **Hunsmann, G., Schneider, J., and Schulz, A.,** Immunoprevention of Friend virus induces erythroleukemia by vaccination with envelope glycoprotein complexes, *Virology,* 113, 602, 1981.

38. **Morein, B., Sundquist, B., Höglund, S., Dalsgaard, D., and Osterhaus, A.,** Iscom, a novel structure for antigenic presentation of membrane proteins from enveloped viruses, *Nature,* 308, 457, 1984.

39. **Osterhaus, A. D. M. E., Weijer, K., UytdeHaag, F., Jarrett, O., Sundquist, B., and Morein, B.,** Induction of protective immune response in cats by vaccination with feline leukemia virus iscom, *J. Immunol.,* 135, 591, 1985.

40. **Sharpe, R. H., Gaulton, G. N., McDade, K. K., Fields, B. N., and Green, M. I.,** Syngeneic monoclonal antiidiotype can induce cellular immunity to reovirus, *J. Exp. Med.,* 160, 1195, 1984.

41. **Greene, M. I., Weiner, H. L., Dichter, M., and Fields, B. N.,** Syngeneic monoclonal anti-idiotypic antibodies identify reovirus type 3 hemagglutinin receptors on immune and neuronal cells, in *Monoclonal and Anti-Idiotypic Antibodies: Probes for Receptor Structure and Function,* Alan R. Liss, New York, 1984, 177.

42. **Ertl, H. C. J. and Finberg, R. W.,** Characteristics and functions of Sendai virus-specific T-cell clones, *J. Virol.,* 50, 425, 1984.

43. **Ertl, H. C. J. and Finberg, R. W.,** Sendai virus-specific T-cell clones: induction of cytolytic T cells by an anti-idiotypic antibody directed against a helper T-cell clone, *Proc. Natl. Acad. Sci.,* 81, 2850, 1984.

44. **Kieber Emmons, T., Ward, R. E., Raychaudhuri, S., Rein, R., and Kohler, H.,** Rational design and application of idiotope vaccines, *Int. Rev. Immunol.,* 1, 1986.

45. **Finberg, R. W. and Ertl, H. C. J.,** Use of T cell-specific anti-idiotypes to immunize against viral infections, *Immunol. Rev.,* 90, 129, 1986.

46. **Stein, K. and Söderstrom, T.,** Neonatal administration of idiotype or anti-idiotype primes for protection against Escherichia coli K13 infection in mice, *J. Exp. Med.,* 160, 1001, 1984.

47. **Hart, D. A., Wang, A. L., Pawlak, L. L., and Nisonoff, A.,** Suppression of idiotypic specificities in adult mice by administration of anti-idiotypic antibody, *J. Exp. Med.,* 135, 1293, 1972.

48. **Eichmann, K.,** Idiotype suppression. I. Influence of the dose and the effector functions of anti-idiotype antibody on the production of an idiotype, *Eur. J. Immunol.,* 4, 296, 1974.

49. **Kennedy, R. C., Adler-Storthz, K., Burns, J. W., Sr., Henkel, R. D., and Dreesman, G. R.,** Antiidiotype modulation of herpes simplex virus infection leading to increased pathogenicity, *J. Virol.,* 50, 951, 1984.

50. **Roehring, J. T., Hunt, A. R., and Mathews, J. H.,** Identification of anti-idiotype antibodies that mimic the neutralization site of Venezuelan equine encephalomyelitis in 1984, Abstract, High Technology Route to Virus Vaccines, Interface Int. Conf., Houston, Tex., 1984, 32.

51. **Kennedy, R. C. and Dreesman, G. R.,** Enhancement of the immune response to hepatitis B surface antigen: in vivo administration of antiidiotype induces anti-HBs that express a similar idiotype, *J. Exp. Med.,* 159, 655, 1984.

52. **Kauffmann, R. S., Noseworthy, J. H., Nepom, J. J., Finberg, R., Fields, B. N., and Greene, M. I.,** Cell receptors for the mammalian reovirus. Monoclonal anti-idiotypic antibody blocks viral binding to cells, *J. Immunol.,* 131, 2539, 1983.

53. **Forni, L., Coutinho, A., Köhler, G., and Jerne, N. K.,** IgM antibodies induce the production of antibodies of the same specificity, *Proc. Natl. Acad. Sci. U.S.A.,* 77, 1125, 1980.

54. **Kang, C. Y. and Kohler, H.,** Immunoglobulin with complementary paratope and idiotope, *J. Exp. Med.,* 163, 787, 1986.

55. **Osterhaus, A. D. M. E. and UytdeHaag, F. G. C. M.,** Lymphocyte hybridomas: production and use of monoclonal antibodies, in *Animal Cell Biotechnology,* Vol. 2, Spier, R. E. and Griffiths, J. B., Eds., Academic Press, London, 1985, 49.

56. **Bunschoten, E. J. et al.,** manuscript in preparation.

57. **Osterhaus, A. D. M. E. et al.,** manuscript in preparation.

Chapter 3

MODULATION OF THE IMMUNE RESPONSE BY ANTI-IDIOTYPIC ANTIBODIES WITH SPECIFICITY FOR B- OR T-CELL IDIOTYPES

H. C. J. Ertl, L. Woo, and R. W. Finberg

TABLE OF CONTENTS

I. INTRODUCTION

The immune system of a vertebrate organism is designed to respond specifically to an indefinite number of foreign molecules. The ability to recognize and eliminate nonself material is of vast importance for the survival of the organism, as it provides the principal defense mechanism against parasites such as bacteria or viruses. Though the ability of the immune system to eliminate foreign material is crucial for the well-being of the host, "pathological" immune responses such as severe allergic reactions or autoimmune responses threaten the very existence of the afflicted host.

Specific modulation of the immune system either by enhancement of wanted immune responses or (alternatively) by suppression of unwanted responses is a major goal of clinical immunologists. Nonspecific modulation, such as by the use of drugs which block protein synthesis or prevent cell division or by the use of antisera which eliminate immunocompetent cells, inevitably leads to increased susceptibility to opportunistic infections or cancer and should, thus, be avoided where more specific remedies are available.

Specific suppression of the immune response can be achieved by some antigens which, at certain high or low doses, are tolerogenic rather than immunogenic. While tolerization can be readily demonstrated with a variety of antigens in neonates, induction of unresponsiveness in an immunologically mature host is often impossible.

Specific stimulation of an immune response can be achieved by vaccination with foreign material. Most vaccines are designed to induce a stage of immunological memory by administration of viruses or bacteria which are immunogenic, but have lost pathogenicity. An optimal vaccine is expected to induce long-lasting immunity of both T and B cells without carrying the risk of reversion to virulence or causing ill effects by other means. Most vaccines which are currently available fail to fulfill these basic requirements: attenuated live vaccines such as vaccinia virus or poliovirus induce long-lasting T- and B-cell immunity, but are potentially hazardous and can cause life-threatening diseases in susceptible individuals. Inert vaccines such as inactivated viruses or microorganisms, subunit vaccines, or synthetic peptides are safer, but are often inefficient in inducing long-lasting immunity. In addition, though these types of vaccines promote production of neutralizing antibodies and induce T cells which mediate a delayed-type hypersensitivity response, they fail to stimulate cytolytic T cells which seem to be crucial for the resistance to many viral infections. With the advent of recombinant DNA technology, engineered recombinant viruses have been made available which induce (like live attenuated viruses) T- and B-cell immunity and might, in some instances, be superior to inert vaccines. Problems with this type of viral vaccines might arise from the viral vector which, in systems using vaccinia or herpesvirus, might be too hazardous for use in humans.

Instead of using foreign material to elicit an immune response, one might modulate the immune system from within, taking advantage of its complementary make-up.

II. NETWORK THEORY

The immune system at a steady stage is thought to be balanced by forming a network of complementary receptors:[1] a portion of the variable region domain of an immunocompetent receptor (the so-called idiotope) is recognized within the immune system by the variable region domain of another receptor (the "anti"-idiotype [anti-Id]). Receptors have more than one idiotope (the sum of which is called an idiotype [Id]) and are, thus, complementary to several anti-Id. The division Id/anti-Id is arbitrary as receptors which are anti-idiotypic in one system, are idiotypic in a different setting, and vice versa. Outside of this autonomous network, portions of the variable region domain of an immunocompetent receptor are able to recognize and respond to foreign molecules (epitopes) and are, in this context, referred to as paratopes. Some idiotopes are identical to paratopes and can, thus, bind epitopes as well as complementary idiotopes of anti-idiotypic receptors. These idiotopes will be referred to in this manuscript as paratactic idiotopes. Other idiotopes are framework associated (regulatory idiotopes). The interaction of these idiotopes with their complementary anti-Id is not inhibitable by conventional foreign antigen.

Antigen competes with the idiotopes of anti-idiotypic receptors for the corresponding paratactic idiotopes and, depending on the affinity of the antigen-paratope vs. the anti-Id - idiotope interaction, disrupts the network. As a consequence, anti-idiotypic receptors are released shortly after administration of antigen as has been demonstrated experimentally.[2] It is not fully understood what precisely triggers the immune response. One could speculate that the epitope-paratope interaction (which has higher avidity as compared to the idiotope-anti-Id interaction) induces expression of receptors for growth factors which leads to proliferation and, in the case of B cells, hypermutation of Id-bearing cells. Upon removal of antigen, an excess of idiotypic receptors causes stimulation of clones with complementary idiotopes. As B cells undergo extensive hypermutation in the course of an immune response, the second set of anti-Id is, presumably, distinct from the first set which balanced the initial Id. Expansion of anti-Id is speculated to down-regulate the idiotypic response until the immune system has reached its steady stage again.

Though in this model anti-Id serve as a balance for Id, administration of an excess amount of certain anti-Id, mainly those which bind to paratactic idiotopes, has consequences similar to administration of antigen, i.e., they induce the immune system to mount an idiotypic (anti-anti-idiotypic) antigen-specific response in the absence of antigen.[3,4]

III. T-CELL ID VS. B-CELL ID

Idiotypic-anti-idiotypic network interactions were first proposed by Jerne[5] for the regulation of antibodies. Abundant experimental evidence has been collected to support this hypothesis. As already mentioned, the specific immune system is composed of two cell types which play distinct roles in the maintenance of self-integrity. B cells which secrete their specific receptors, i.e., antibodies, as effector molecules, recognize antigen with high affinity and are teleologically presumed to police the extracellular compartment of an organism by binding soluble foreign material. T-cell receptors fail to recognize soluble foreign antigen, but require corecognition of self-cell-surface markers which are encoded for by genes of the major histocompatibility complex (MHC).[6,7] HLA in men and H-2 in mice. They are, thus, optimally equipped for surveillance of the cellular compartment of an organism without being susceptible to paralysis by soluble antigen.

T- and B-cell receptors are encoded for by different genes which evolved from a common primordial cell surface protein.[8] Both receptors consist of two chains, heavy and light chain for immunoglobulins and α and β chain for T-cell receptors, which are linked by disulfide bridges. Both receptors have highly conserved (constant) regions and variable regions which determine the antigen specificity. Though both receptors share these common features, the fine structure of the variable regions is quite distinct. The antigen (and anti-Id) binding regions of several antibodies have been characterized by peptide and/or cDNA sequencing[9] which, together with functional data and crystallographic analysis,[10] have provided us with a fundamental understanding of antigen/Id and Id/anti-Id interactions.

The T-cell receptor has been far more elusive. Though both chains of this receptor are by now well characterized at a molecular level,[11,12] the mechanism of the antigen-MHC molecule recognition is by no means understood. Several models which are in agreement with our limited knowledge are currently being tested. The models are based on the requirement of T-cell receptors to bind to cell-bound antigen without being blocked by soluble antigen. The affinity for antigen has been shown to be sufficiently low to avoid this, since binding only occurs if antigen is associated with glycoproteins encoded for by the MHC gene complex. Recognition of self-cell-surface molecules has to be strictly controlled to prevent autoimmune response, which means that any direct interaction between T-cell receptor and MHC antigens has to be of an affinity too low to trigger a response independently.

Let us briefly describe three different models of T-cell receptor/antigen-MHC antigen recognition:

1. Two separate receptor molecules, one with low affinity for MHC molecules and one with low affinity for foreign antigen, might meet the above-described requirements, but none of the currently available data support this dual receptor model.
2. Two separate binding sites on one receptor which interact with antigen and MHC molecules. In this model MHC antigen and foreign antigen can be either recognized as independent entities or, more likely, as recent studies from two laboratories suggest,[13,14] form a trimolecular complex with the T-cell receptor. In a trimolecular complex three sites of interactions, i.e., T-cell receptor/antigen, antigen/MHC molecule, and T-cell receptor/MHC molecule, determine the final strength of the T-cell/target-cell interaction which would not be stable if one of the components (MHC molecule for soluble antigen, antigen for normal tissue cells) was missing.
3. Another model, the altered self-hypothesis, postulated in 1974 by Zinkernagel and Doherty[6] to explain H-2 restriction, presumes a single receptor site for an H-2 antigen which has been modified by antigen.

Serological and cellular studies have shown cross reactivity between T-cell receptors and immunoglobulins,[15-17] indicating shared Id on T and B cells. The T-cell receptor was initially isolated by serological means,[18] suggesting that B cells can mount an anti-idiotypic response to T-cell receptors. Conversely, T cells have been shown to recognize B-cell Id or anti-Id.[19,20] It, thus, seems likely that T- and B-cell receptors have sufficient complementarity (not necessarily at a molecular level, but at a structural level) to either participate in one network or, alternatively, to form separate but overlapping networks.

In a common network, T-cell idiotopes would be expected to resemble B-cell idiotopes.

In a dual receptor model or a dual binding site model for T-cell recognition (the altered self-model is not compatible with a single or overlapping network between T-cell idiotopes and — at least — paratactic B-cell idiotopes; less is known about regulatory B-cell idiotopes which might conceivably have complementary to antiself-receptors), a B cell-T cell network would, presumably, utilize the antigen binding site of the T-cell receptor (there is no equivalent of MHC restriction governing antibody-antigen interactions). There is some experimental evidence for this speculation.

Anti-Id specific for immunoglobulin Id have been shown to stimulate antigen-specific T cells.[19,20] One set of data indicated that this interaction is governed by H-2 restriction.[20] In this experimental system, the anti-Id, presumably, mimic an epitope which is recognized both by B cells and the antigen binding site of the T-cell receptor.

Anti-idiotypic antibodies with specificity for the T-cell receptor have been shown in vitro to stimulate (upon addition of lymphokines) T cells without requiring corecognition of H-2 molecules.[21] In vivo, T-cell-defined antibodies were shown to induce an idiotypic (i.e.) antigen-specific T-cell response in hosts which differed in their H-2 haplotype.[22]

Idiotypic antibodies have been shown to induce (anti-idiotypic) T cells, which regulate expression of the Id in association with a certain isotype, indicating that T-cell-mediated network control can follow a different pathway as compared to antigen-driven T-cell responses.[23]

T suppressor cells, which are fundamentally distinct from effector T cells as shown by their ability to bind soluble antigen and, accordingly, their lack of H-2 restriction, have been shown to induce a cascade of anti-idiotypic-idiotypic cells and factors which cross react with B-cell-defined idiotopes.[24,25]

IV. USE OF ANTI-ID AS VACCINES

As described above, anti-Id can be divided into several groups by their binding specificity and function. Since this division is important for the choice of an anti-Id vaccine, let us briefly summarize: anti-Id can be generated against B cells (referred to as B-cell-defined Id or T cells (T-cell-defined anti-Id). Anti-Id to B cells can recognize the antigen binding site of an immunoglobulin molecule (i.e., a paratactic idiotope), some of those can mimic antigen, i.e., carry an internal image of the antigen, and other B-cell-defined anti-Id recognize framework or regulatory idiotopes of the antibody molecule which are not associated with the antigen binding sites. The fine specficity of T-cell-defined anti-idiotypic antibodies which has not yet been dissected might follow a similar pattern.

Anti-Id vaccines would have to fulfill the same basic requirements as mentioned earlier for conventional vaccines, i.e., they would have to be free of ill effects and they would have to be able to induce B- as well as T-cell immunity involving cytolytic T cells and T cells which mediate a delayed-type hypersensitivity response. Cell-bound anti-Id, such as T-cell receptors, are inacceptable for use in humans. Extensive purification or synthesis in appropriate vectors would be required to make the use of these anti-Id more feasible. Soluble monoclonal anti-idiotypic antibodies could be easily produced in large quantities. The risk of adverse reactions due to allotype differences could be overcome by the generation of chimeric antibodies[26] between murine monoclonal anti-Id and human immunoglobulins. Other adverse reactions due to suppression rather than stimulation of an idiotypic immune response or due to complementarity to structures other than immunocompetent receptors (such as drug or hormone receptors[27,28]) would have to be excluded for every anti-Id.

A. B-Cell-Defined Anti-Id

A number of investigators have used B-cell-defined anti-idiotypic antibodies to induce immunity to microorganisms, viruses, or transformed cells. Let us briefly review some of their work.

1. Induction of Immunity to Parasitic Infections

a. Trypanosoma rhodesiense

A polyclonal anti-idiotypic antiserum to a panel of monoclonal idiotypic antibodies with specificity to *T. rhodesiense* was shown to protect mice against a subsequent infection with parasite. Protection was presumably caused by the induction of neutralizing antibodies.[29] The anti-Id did not mimic antigen, since induction of the response was genetically restricted.[30]

b. Schistosoma mansonii

A monoclonal IgM antibody was induced against a monoclonal IgG_2 antibody which was cytotoxic for parasites in presence of eosinophiles and upon passive transfer protected rats against a lethal infection. The anti-Id induced antigen-specific cytotoxic antibodies. A high percentage of rats which had been vaccinated with the anti-Id was protected against an infectious challenge.[31]

2. Induction of Immunity to Viral Infections

a. Hepatitis B Virus

A xenogeneic anti-Id serum to a common idiotope of hepatitis B virus-specific antibodies was used to induce an antibody response. This anti-Id serum contained antibodies which, presumably, expressed an internal image of a hepatitis B virus epitope, as a response could be elicited in several species.[32] Whether this anti-Id antiserum induced T cells was not investigated.

b. Rabies Virus

Several anti-idiotypic antibodies to rabies virus-specific monoclonal antibodies were shown to induce neutralizing antibodies in mice. The amount of neutralizing antibodies was too low to protect against a lethal infection, but could be boosted by a subsequent subimmunogeneic dose of viral antigen.[33] Rabies virus infection is characterized by an exceptionally long incubation time in humans (but not in mice). In most cases, the disease is prevented by postexposure vaccination. Preimmunization with a well-tolerated anti-Id might be an alternative to the currently available vaccines. Another interesting finding in this study was that the specificity of the anti-Id-induced antibody response was not necessarily identical to the Id used for generation of the anti-Id: one of the anti-idiotypic sera which was directed to a nonneutralizing monoclonal antibody was found to induce a neutralizing antibody response.

c. Poliovirus Type II

One group produced a monoclonal anti-idiotypic antibody to a neutralizing antibody specific for poliovirus type II. The anti-Id induced neutralizing antibodies, but failed to protect mice against a subsequent lethal infection.[34]

d. Venezuelan Equine Encephalitis Virus

A cross-species anti-idiotypic serum to monoclonal antibodies specific for the neutralizing site of Venezuelan equinine encephalitis virus was shown to induce an antiviral antibody response in mice.[35] An anti-Id vaccine in this system might be of particular value as conventional vaccines were shown to cause disease in humans.[36]

e. Tobacco Mosaic Virus

Francotte and Urbain[37] used a rabbit-derived anti-idiotypic serum to a private idiotope to induce antitobacco mosaic virus-specific antibodies in mice. Comparing anti-Id-induced antibodies with antigen-induced antibodies, it was found that the anti-Id stimulated silent B-cell clones which do not participate in the immune response upon inoculation of nominal antigen. The anti-Id was, thus, paratope specific without carrying the internal image of the epitope.

f. Reovirus Type III

A monoclonal antibody to the neutralizing site of the reovirus type III hemagglutinin molecule was used to generate a monoclonal anti-idiotypic antibody of the IgM class. This monoclonal antibody was found to induce neutralizing antibodies and T cells which, upon adoptive transfer, mediated a delayed-type hypersensitivity response to reovirus.[20] Immunization with soluble antibody did not induce cytolytic T cells, though injection of anti-Id expressing hybridoma cells did, thus, presumably, reflecting a requirement for corepresentation with MHC-encoded determinants. In vitro anti-Id-expressing hybridoma cells served as targets for reovirus-specific cytolytic T cells.[38,39] In addition to its immunological functions, the anti-Id was found to compete with reovirus for its cell surface receptor,[28] which could conceivably be useful therapeutically.

3. Induction of Immunity to Bacterial Infections
a. Streptococcus pneumonia

Using a monoclonal anti-Id to a phosphorylcholine-specific antibody, one group described induction of protective immunity to a subsequent lethal *S. pneumonia* infection in mice. To enhance immunogenicity, the anti-Id was coupled to a carrier protein. Protection was achieved by the stimulation of antibacterial antibodies.[40]

b. Escherichia coli

A very elegant report describes the use of an anti-Id in neonates. Newborn infants fail to mount an immune response to polysaccharide molecules such as the capsular antigens of *E. coli*. Stein et al. demonstrated that inoculation of neonatal mice with an anti-Id specific for an antipolysaccharide antibody enabled these mice to mount a protective immune response to a subsequent inoculation of bacteria.[41] Anti-Id which mimic a polysaccharide molecule are, thus, able to overcome a stage of unresponsiveness which a conventional (nonprotein) vaccine would not achieve.

4. Induction of Immunity to Tumor Antigens
a. Adenocarcinoma

Gorczynski and co-workers[42] described experiments which indicated that immunization of mice with anti-idiotypic serum to monoclonal antibodies specific for embryo-associated antigenic determinants inhibited the growth of subsequently transplanted adenocarcinoma cells. Immunization of mice with monoclonal idiotypic antibodies, on the other hand, enhanced tumor growth, presumably, by selection of antigenic variants. These data indicate that anti-Id can cause protection other than by the induction of idiotypic antibodies.

b. Melanoma

Nepom et al.[43] raised anti-idiotypic antibodies in rabbits to a murine monoclonal antibody specific for the human melanoma-associated cell surface antigen p97. Mice injected with anti-Id mounted a transferable delayed-type hypersensitivity response to tumor cells as well as anti-anti-idiotypic antibodies, some of which recognized the original antigen.

c. SV-40-Induced Sarcomas

Four idiotypic antibodies directed to Simian virus 40 tumor antigen or the cellular protein p53 were used to induce anti-idiotypic antisera. Sixty percent of mice which were immunized with pooled anti-idiotypic sera showed partial protection of tumor formation upon inoculation of SV-40 transformed cells. Suppression was probably not due to generation of tumoricidal antibodies, since mice which had been immunized with anti-Id developed anti-anti-idiotypic antibodies which failed to recognize SV 40 tumor antigen or p53.[44]

5. Induction of Increased Susceptibility
a. Herpes Simplex Virus Type I

Systems described so far dealt with the use of anti-Id to induce immunity, which either successfully provided protection or had no effect on the outcome of the subsequently induced disease. One report has been published which describes increased susceptibility of anti-Id-primed mice to a subsequent inoculation of herpex simplex virus.[45] The mechanism(s) of this important finding was not further investigated.

To summarize, the above-described data clearly indicate that B-cell-defined anti-Id can be used to induce protective immunity to parasites, viruses, bacteria, or transformed cells (Table 1). Anti-Id which carry the internal image of an antigen are excellent candidates for vaccines, since they induce antigen-specific antibodies and, as shown in two systems,[20,43] T cells which mediate a delayed-type hypersensitivity response. In some virus infections, anti-Id might be developed which bind virus receptors and can, thus, be used to inhibit the entry of virus into cells if given shortly after exposure to antigen.

Unfortunately, B-cell-defined soluble anti-idiotypic antibodies seem to be unable to stimulate Class I restricted, cytolytic T cells which are crucial for the resistance to many viral infections.[20] As B-cell-defined anti-Id have been shown in numerous systems to induce Class II region-restricted T cells (i.e., T cells which mediate a delayed-type hypersentitivity response and T helper cells) and Class I and II-restricted T cells utilize the same genetic material, it seems unconceivable that the lack of stimulation of a cytolytic T-cell response is due to differences in the genetically determined repertoire. It is more likely that cytolytic T cells require a different type of antigen presentation which is not provided for by anti-idiotypic antibodies (or inert particles, in general[46]). Data obtained in the reovirus system confirm this as anti-Id-presenting hybridoma cells, but not soluble antibodies, induced an antigen-specific cytolytic T-cell response.[29] Though the induction of cytolytic T cells by B-cell-defined anti-idiotypic antibodies has to be tested in other systems, the limited amount of data available so far indicates that B-cell-defined anti-idiotypic antibodies which carry an internal image act like inert vaccines.

Anti-Id specific for a paratopic idiotope which does not mimic antigen exist at a much higher frequency as compared to anti-Id with an internal image.[47] This type of antibodies has been used successfully to induce immunity to microorganisms such as *Trypanosoma*.[29,30] Unfortunately, in this system a response could only be induced in hosts of a certain genetic haplotype. Genetic restriction might be an intrinsic problem for the use of paratope-specific anti-Id (which do not present an internal image) as they, presumably, induce a very limited number of B-cell clones with the desired antigen specificity. This type of antibody will primarily stimulate nonantigen-specific anti-anti-idiotypic B-cell clones which are irrelevant to protection against a given infectious reagent. In some systems, though, paratope-specific anti-Id might induce silent B-cell clones[37] (i.e., B-cell clones which would not have been activated upon exposure to antigen, but which are, nevertheless, antigen specific) and, thus, stimulate an immune response across species barriers.

In some instances, it might be an advantage to suppress rather than to stimulate an immune response. Large doses of anti-idiotypic antibodies administered at birth or in adults can cause long-lasting Id suppression.[48] Some infectious diseases are complicated in susceptible individuals by autoimmune responses. For example, Epstein-Barr virus infections are commonly followed by antibody-mediated hemolytic anemia and several viral infections are associated with autoimmune encephalitis. Upon identification of the autoreactive Id, an anti-Id given at a tolerable dose might prevent these complications.

Table 1
USE OF ANTI-ID AS IMMUNOGEN

Antigen	Anti-Id	Immune response	Ref.
Parasites			
Trypanosoma rhodesiense	B.c.d., xenogeneic polyclonal serum, paratope specific	Induction of neutralizing antibodies, protective immunity, genetically restricted	29, 30
Schistosoma mansonii	B.c.d., monoclonal IgM, internal image	Induction of cytotoxic antibodies, protective immunity	31
Viruses			
Hepatitis B virus	B.c.d., xenogeneic polyclonal serum, internal image	Induction of neutralizing antibodies in several species, T cells???	32
Rabies virus	B.c.d., monoclonal, paratope specific	Induction of neutralizing antibodies, incomplete protection, T cells???	33
Poliovirus type II	B.c.d., monoclonal, internal image	Induction of neutralizing antibodies, no protection, T cells???	34
Venezuela equine encephalitis virus	B.c.d., xenogeneic polyclonal serum, internal image	Induction of neutralizing antibodies, T cells???	35
Tobacco mosaic virus	B.c.d., xenogeneic polyclonal serum, paratope specific	Stimulation of silent B-cell clones, T cells???	37
Reovirus type III	B.c.d., syngeneic monoclonal IgM, internal image	Induction of neutralizing antibodies, T cells which mediate a delayed-type hypersensitivity response and cytolytic T cells, binding to the virus receptor	20
Herpes simplex virus	B.c.d., xenogeneic polyclonal serum, paratope specific?	Increased susceptibility to viral challenge	44
Sendai virus	T.c.d., monoclonal syngeneic IgM, paratope specific?	Induction of cytolytic T cells and T cells which mediate a DTH response, induction of antibodies, protective immunity	53
Bacteria			
Streptococcus pneumonia	B.c.d., monoclonal IgG, internal image	Induction of neutralizing antibodies, protective immunity	40
Escherichia coli	B.c.d., monoclonal IgG1, internal image?	Induction of neutralizing antibodies and protective immunity in neonates	41
Listeria monocytogenes	T.c.d., polyclonal syngeneic serum	Induction of protective immunity	54
Tumors			
Adenocarcinoma	B.c.d., polyclonal xenogeneic serum, paratope specific?	Suppression of tumor growth	42
Melanoma	B.c.d., xenogeneic polyclonal serum, paratope specific?	Induction of T cells which mediate a DTH response, induction of an anti-anti-Id B-cell response, in some mice induction of tumor-specific antibodies	43
SV-40 induced sarcoma	B.c.d., xenogeneic polyclonal serum, paratope specific?	Partial suppression of tumor growth, induction of anti-anti-Id, no induction of tumor-specific antibodies	44

Note: B.c.d., B-cell-defined anti-Id antibodies.
T.c.d., T-cell-defined anti-Id antibodies.

B. T-Cell-Defined Anti-Id

Much less data have been collected on the use of T-cell-defined anti-idiotypic reagents as immunogens. Using mixed populations of alloreactive T cells, Frischknecht and co-workers[49] prepared anti-idiotypic antisera which were shown to substitute for histocompatibility antigens and stimulate allo-antibody synthesis as well as cytolytic T lymphocytes in vivo. These data clearly indicated that T-cell-defined anti-idiotypic antbodies could induce a B- and T-cell response.

Subsequently, with the identification of soluble T-cell growth factors,[50] techniques for long-term culture of pure T-cell lines were developed.[51] Using cloned T-cell lines or hybridomas, anti-idiotypic antibodies were generated which were shown to precipitate the T-cell receptor, a -80 to 90,000-dalton dimer,[18] which is by now well characterized. Anti-Id to T-cell hybridomas were shown to suppress or (upon aggregation) induce an antigen-driven response of the appropriate T-cell clones in vitro. As we described above, B-cell-defined anti-idiotypic antibodies seem to be unable to induce a cytolytic T-cell response. The use of T-cell-defined anti-Id might provide an alternative in systems where T-cell-mediated immunity is crucial.

1. Sendai Virus

To address this question, we used Sendai virus as antigen. Sendai virus, a parainfluenza virus type I, is a natural mouse pathogen which causes upon intranasal inoculation an often fatal pneumonia. Different mouse stains differ in their susceptibility to Sendai virus, and though the genetic factor(s) which transfer resistance are not yet fully known, an increase in susceptibility is linked to a lack of functional T cells. Nude mice are more sensitive to lethal infection than normal littermates.

a. Characteristics of the Parental T-Cell Clone

Assuming that helper T cells and cytolytic T cells might utilize the same genetic repertoire, we used a Sendai virus-specific T helper cell clone to generate an anti-T-cell antibody. The T-cell clone (termed 2H3.E8[51]) was of B10.D2 origin. 2H3.E8 was Thy 1.2$^+$, Ly-1.2$^-$, and Lyt-2.2$^-$. The clone proliferated and secreted lymphokines upon coculture with I region-compatible, virus-presenting stimulator cells, but failed to lyse virus-infected target cells. In vivo, the injection of 2H3.E8 into syngeneic mice induces a delayed-type hypersensitivity response to Sendai virus and under appropriate experimental conditions augmented a cytolytic T-cell response to a subimmunogenic dose of Sendai virus. This clone represents, thus, a typical antigen-specific T helper cell clone.

b. Generation of a Monoclonal Anti-T-Cell Antibody

The 2H3.E8 T-cell clone was injected over a period of 6 months in 14-day intervals into syngeneic B10.D2 mice. Splenocytes from the immunized mice were subsequently fused with NS-1 myeloma cells and hybrid supernatants were tested for binding to the 2H3.E8 T cell clone by indirect immunofluorescence and subsequent analysis in a fluorescence-activated cell sorter. One hybrid was found to be positive. Two subclones, termed 1B4.E6 and 1B4.H6, which secreted T-cell-specific antibodies, were further investigated.

c. Characterization of the Anti-T-Cell Antibody

The 1B4.E6/H6 hyridoma line secreted an IgM antibody which was shown to be specific for an idiotope present on a high percentage of Sendai virus-specific T-cell clones. As the antibody failed to exhibit binding to naive T cells or T cells directed to irrelevant antigens, we assume that the 1B4.E6/H6 antibody recognizes a cross-reactive idiotope of the Sendai virus-specific T-cell repertoire (Table 2). Its anti-Id specificity was confirmed by its in vitro functions, i.e., upon coculture with appropriate stimulator cells the anti-Id induced prolif-

Table 2
SPECIFICITY OF 1B4.E6/H6

Sendai virus-specific T-cell clones	30—40% Positive
Reovirus-specific T-cell clones	Negative
1B4.E6/H6-induced T-cell clones	Positive
Sendai virus-induced T-cell populations	3—15% Positive
1B4.E6/H6-induced T-cell populations	3—15% Positive
Ly-1$^+$ Sendai virus-specific T-cell populations	3—15% Positive
Lyt-2$^+$ Sendai virus-specific T-cell populations	<2% Positive
Alloantigen-induced T-cell populations	Negative
Reovirus-induced T-cell populations	Negative
Naive splenocytes	Negative
Thymocytes	Negative

Note: Tested by indirect immunofluorescence and analysis in a fluorescence-activated cell sorter.

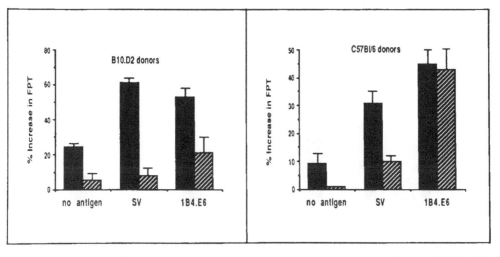

FIGURE 1.   Induction of T cells which mediate a delayed-type hypersensitivity response. Groups of B10.D2 (H-2^d) or C57B1/7 (H-2^b) mice (i.e., donor mice) were injected with either 10^3 hemagglutinating units of UV-light inactivated Sendai virus or 200 μℓ of 1B4.E6 culture supernatant. Splenocytes of immune and naive (no antigen) mice were harvested 7 days later and transferred i.v. into naive B10.D2 (solid bar) or C57B1/6 (hatched bar) mice, which were challenged with 10^3 HAU infections Sendai virus into the left footpad. Footpad thickness was measured 24 hr later. Percent increase in footpad thickness was calculated using the formula:

$$\frac{\text{FPT left foot - FPT right foot}}{\text{FPT left foot}} \times 100$$

eration of its parental T-cell clone (data not shown). In addition, the 1B4.E6/H6 hybridoma line served as a target for Sendai virus-specific T cells (see later), thus, indicating that the 1B4.E6/H6 idiotope was expressed both on Class I- and II-restricted T cells.

d. *In Vivo Functions of the Anti T-Cell Antibody*
i. Induction of T cells Which Mediate a Delayed-Type Hypersensitivity Response[53]

Mice which were injected with 1B4.E6/H6 culture supernatant generated T lymphocytes, which, upon adoptive transfer, mediated a delayed-type hypersensitivity response in vivo. In contrast to the response generated by Sendai virus, the anti-Id-induced response was characterized by an apparent lack of H-2 restriction (Figure 1). The response could be evoked in mice of any H-2 haplotype (Figure 1) or IgH allotype (Figure 2) tested so far.

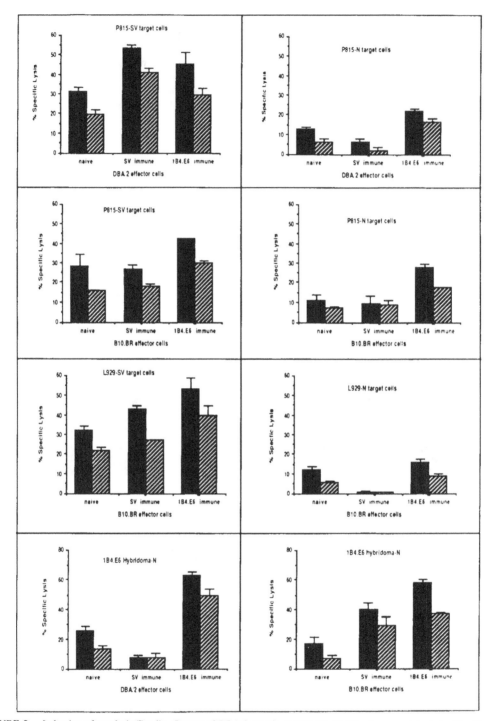

FIGURE 2.   Induction of cytolytic T cells. Groups of DBA/2 (H-2^d) or B10.BR (H-2^k) mice were injected with either 10^3 hemagglutinating units of UV-light inactivated Sendai virus or 200 μℓ of 1B4.E6 culture supernatant. Splenocytes of immune and naive mice were harvested 7 days later and restimulated in bulk culture for 5 days with UV-light inactivated Sendai virus. Effector cells were harvested, purified over a ficoll-gradient, and tested for lysis of uninfected and Sendai virus-infected P815 (H-2^d) and L-929 (H-2^k) target cells and uninfected 1B4.E6 hybridoma cells in a 4-hr ^{51}CR-release assay at effector-to-target cell ratios of 20:1 (solid bar) and 10:1 (hatched bar). Percent specific lysis was calculated using the formula:

$$\frac{\text{Cr release in presence of effector cells} - \text{Cr release in presence of medium}}{\text{Maximal Cr-release} - \text{Cr release in presence of medium}} \times 100$$

ii. Induction of Cytolytic T Cells[52]

Immunization of mice with 1B4.E6/H6 supernatant induced Lyt-2 $^+$ T lymphocytes which, upon restimulation in vitro (with UV-light inactivated Sendai virus), lysed antigen-presenting target cells (Figure 2). A response could be evoked in mice of different strains. Though the response was, in most cases, lower as compared to a virus-induced response, H-2^k mice which are low responders to Sendai virus showed higher lysis of Sendai virus-infected target cells upon injection of the antibody as compared to virus. As described above for the delayed-type hypersensitivity response, lysis was mediated across H-2 barriers; 1B4.E6 immune H-2^k effector cells lysed H-2 compatible infected target cells (L-929) as well as incompatible infected tumor target cells (P815) (Figure 2). Using limiting dilution conditions (data not shown), we were able to demonstrate that up to 60% of clones lysed allogeneic infected target cells as well as syngeneic infected target cells.

As had been shown for the immune response to reovirus, the 1B4.E6 hybridoma cell line, which expresses the anti-Id as a surface immunoglobulin, was recognized by cytolytic T cells induced by anti-idiotypic antibody or, as shown for B10.BR effector cells, by viral antigen. Recognition of 1B4.E6, even by Sendai virus-induced effector cells (which are H-2 restricted), did not require association of the antibody with compatible H-2 antigen. To test if the same receptor is used to recognize antigen + H-2 or anti-Id (without H-2?), we generated a T-cell clone from a Sendai virus-immune B10.D2 mouse which lysed Sendai virus-infected target cells as well as the 1B4.E6/H6 hybridoma cell line. The clone which was characterized by indirect immunofluorescence and subsequent analysis in a fluorescence-activated cell sorter as being Lyt-1.2$^-$, Lyt-2.2$^+$, L3T4$^-$, Thy 1.2$^+$, and 1B4.E6/H6$^+$ failed to lyse unmodified tumor target cells or myeloma cells, thus, indicating that lysis of the hybridoma cell was "antigen" specific and not caused by a nonspecific natural killer cell-like activity (data not shown). Lysis of the hybridoma cell could be inhibited by Sendai virus-infected target cells and vice versa, thus, indicating that lysis was mediated by the same cell (and not two separate populations which might have evolved during in vitro culture) and, presumably, by the same receptor. The hybridoma line was more efficient than the virus-infected tumor cell in inhibiting both lysis of the infected target cells or the uninfected 1B4.H6 hybridoma cell (Figure 3).

Though Sendai virus-specific cytolytic T cells could be easily induced in vivo with the 1B4.E6/H6 antibody upon in vitro restimulation with Sendai virus, we were unable to obtain virus-specific killer cells by restimulating 1B4.E6 immune splenocytes in vitro with antibody instead of antigen. Killer cells obtained under these conditions were completely nonspecific and lysed uninfected target cells and Sendai virus-infected target cells equally well (Figure 4). It is unlikely that this is caused by inappropriate presentation of the anti-Id in vitro, as Sendai virus-immune splenocytes which were restimulated in vitro with antibody showed a small percentage of specific lysis on infected target cells. It is more likely that 1B4.E6 induces a variety of immune cells, most of which fail to recognize Sendai virus-infected target cells. In vitro restimulation with virus might be necessary to expand the small proportion of virus-specific 1B4.E6/H6$^+$ T-cell clones which, upon restimulation with antibody, might be outnumbered and overgrown by cells which are anti-anti-idiotypic, but lack antigen specificity.

The experiments we presented so far dealt with the induction of an acute response to Sendai virus. To test if the 1B4.E6 antibody induces a long-lasting memory response, mice were injected either with Sendai virus or various concentrations of purified 1B4.H6 antibody. Three months later, splenocytes were restimulated in vitro with virus either under limiting dilution conditions or in bulk culture. Even after 3 months, which is nearly an $^1/_8$th of an average mouse life, 1B4.H6-primed splenocytes mounted a response to Sendai virus. The response was strictly dose dependent, i.e., splenocytes from mice which obtained the highest dose of antibody (i.e., 0.5 mg) had a higher frequency of cytolytic T cells as compared to mice which had been immunized with an intermediate (5 μg) or a low (50 ng) dose of antibody (Table 3).

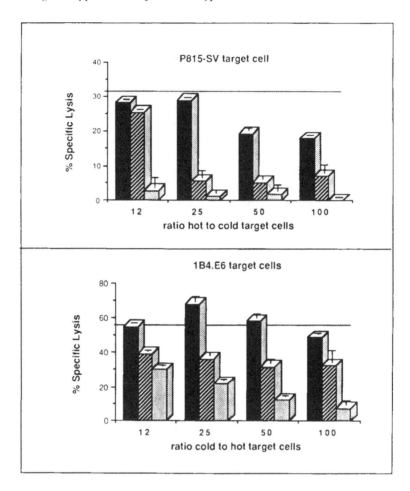

FIGURE 3.   Induction of cytolytic T cells: cold target cell inhibition. T-cell clone 2H10.G8 was tested for lysis of SV-infected P815 cells (effector-to-target cell ratio 40:1) and uninfected 1B4.E6 hybridoma cells (effector-to-target cell ratio 10:1). Unlabeled target cells were added at various concentrations: (solid bar) cold-uninfected P815 tumor cells, (hatched bar) cold Sendai virus-infected P815 tumor cells, (cross-hatched bar) uninfected 1B4.E6 hybridoma cells. Control lysis in absence of cold target cells was 38% for Sendai virus-infected P815 target cells and 58% for 1B4.E6 target cells (−).

iii. Induction of Proliferative T-Cell Clones

To test if antigen-induced T cells and anti-Id-induced T cells exhibit identical or distinct specificities, T-cell clones were generated from splenocytes of B10.D2 mice which had been immunized in vivo with 1B4.E6/H6 and, subsequently, restimulated in vitro with viral antigen. Clones derived from several different experiments were carried on Sendai virus-presenting syngeneic splenocytes derived from mice which had been immunized 2 to 4 months previously with Sendai virus, as had been described for antigen-induced T-cell clones.[52] The 1B4.E6/H6-induced T-cell clones were phenotypically undistinguishable from antigen-induced T helper cell clones i.e., they were Thy-1.2$^+$, Ly-1.2$^+$, Lyt-2.2$^-$, L3T4$^+$, and 1B4.E6/H6$^+$. Anti-Id-induced T cell clones (such as 2D3.F1, Figure 5, or ID.B7.A8, Figure 6) showed consistently high proliferation upon coculture with splenocytes of the H-2^d or H-2^b (but not H-2^k) haplotype in absence of viral antigen. Pretreatment of B10.D2 or C57B1/6 splenocytes with Sendai virus augmented proliferation of 1B4.E6/H6-induced T cells only marginally. In most experiments a significant proliferative response to Sendai

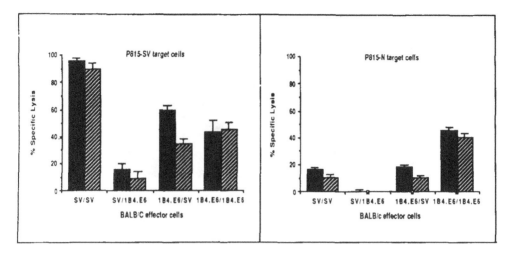

FIGURE 4. Induction of cytolytic T cells. BALB/c mice were primed with 10³ HAU Sendai virus or 200 µℓ of 1B4.E6 culture supernatant. Seven days later splenocytes were restimulated with either 1B4.E6 culture supernatant or UV-light inactivated Sendai virus. Cytolytic activity of effector cells was tested 5 days later in a ⁵¹Cr-release assay on uninfected or Sendai virus-infected P815 target cells at responder-to-target cell ratios of 40:1 (solid bar) and 20:1 (hatched bar).

Table 3
INDUCTION OF A MEMORY RESPONSE

Effector cells	Specific lysis ± SD of P815 SV (P815N) (%)	
	40:1	**5:1**
DBA/2N/SV	22 ± 2 (18 ± 1)	6 ± 2 (0.1 ± 1)
DBA/2SV/SV	51 ± 2 (35 ± 5)	13 ± 3 (3 ± 2)
DBA/2 500 µg Ab	37 ± 3 (25 ± 4)	9 ± 1 (4 ± 1)
DBA/2 5 µg Ab	31 ± 2 (19 ± 1)	4 ± 1 (1 ± 2)
DBA/20.05 µg Ab	28 ± 2 (19 ± 3)	4 ± 2 (1 ± 3)

Note: Groups of DBA/2 mice were immunized with 10³ HAU UV-light inactivated Sendai virus or various concentrations of purified 1B4.H6 antibody. Three months later, splenocytes were restimulated in vitro for 5 days with inactivated Sendai virus. Effector cells were tested at effector-to-target cell ratios of 40:1 and 5:1 on Sendai virus-infected and uninfected (bracketed values) P815 target cells in a 4-hr ⁵¹Cr-release assay.

virus (>2× the proliferation to unmodified cells) was only observed upon coculture with Sendai virus-presenting DBA/2 splenocytes which caused a higher degree of antigen-specific proliferation of Sendai virus-induced clones (for example, SV.G5.E8, shown in Figure 6). Antigen-specific proliferation of anti-Id-induced T cells (but not of Sendai virus-induced T-cell clones) was significantly augmented by presentation with Sendai virus on splenocytes of mice which had been immunized several weeks previously with Sendai virus. This was probably not due to lymphokines secreted by Sendai virus-immune T cells present in the stimulator cell population, since the IL-2-dependent HT-2 T-cell line which was tested in parallel failed to proliferate under these conditions. We assume that antigen-presenting cells derived from immune spleen (and to a lesser extent derived from DBA/2 spleens) might

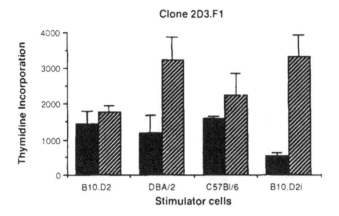

FIGURE 5.   Specificity of 1B4.H6-induced T-cell clones. Clone 2D3.F1
$(2 \times 10^4$/well) was cocultured with unmodified (solid bar) or Sendai virus
pretreated (hatched bar) gamma-irradiated splenocytes $(1 \times 10^6$/well) of
naive B10.D2, DBA/2, or C57B1/6 mice or splenocytes from B10.D2
mice which had been immunized with UV-light inactivated Sendai virus
2 to 4 months previously (B10.D2i). Proliferation was tested 48 hr later
by an 8-hr ^{3}H-thymidine pulse.

present virus more efficiently as compared to naive splenocytes and might, thus, facilitate
proliferation of T-cell clones with low affinity to Sendai virus. None of the anti-Id-induced
T-cell clones secreted measurable amounts of IL-2 upon coculture with Sendai virus-presenting stimulator cells. Their function is, thus, unclear.

iv. Induction of an Antibody Response

To test if 1B4.E6/H6 induced a virus-specific T-cell response, sera of mice which had
been inoculated with antibody were screened for antiviral immunoglobulin. Sendai virus-
specific antibodies could be found in several mouse strains (one example shown in Figure
7). Though the antibody response could be elicited in every mouse strain we tested, mice
of B10 backgroun showed, in general, low responses as compared to mice of other backgrounds, which might indicate that IgH differences enhance B-cell stimulation by providing
a carrier effect. The antibody response was under T-cell control, i.e., nude mice neither
generated antibodies to Sendai virus nor to 1B4.E6/H6. To test if anti-Id-induced antibodies
were anti-anti-idiotypic, i.e., bound the 1B4.E6/H6 antibody, sera of 1B4.H6-immune mice
or as a control of Sendai virus-immune mice were adsorbed to CNBr-activated Sepharose®
beads which had been coupled with purified 1B4.H6 antibody. When the nonadsorbed
fraction and the adsorbed fraction (eluate) were tested for Sendai virus specificity, only a
small percentage of the 1B4.H6-induced Sendai virus-specific immunoglobulins bound to
1B4.H6 (Figure 8). The highest percentage of antibodies specific for Sendai virus *and* 1B4.H6
was found in the IgG fraction of mice which had been immunized 42 days previously with
anti-Id.

e. Mode of Action of the Anti-T-Cell Antibody

The above-presented data show clearly that the 1B4.E6/H6 antibody can induce in vivo
a Sendai virus-specific immune response which, as been reported previously, can protect
mice against a subsequent infection with a lethal dose of Sendai virus.[53]. These data do not
yet allow us to deduce the exact mechanisms by which the anti-T-cell antibody induces an
immune response. Several pathways could be envisioned:

1.   **Mimicry of antigen (Sendai virus):** Does 1B4.E6/H6 mimic antigen? Antigenic

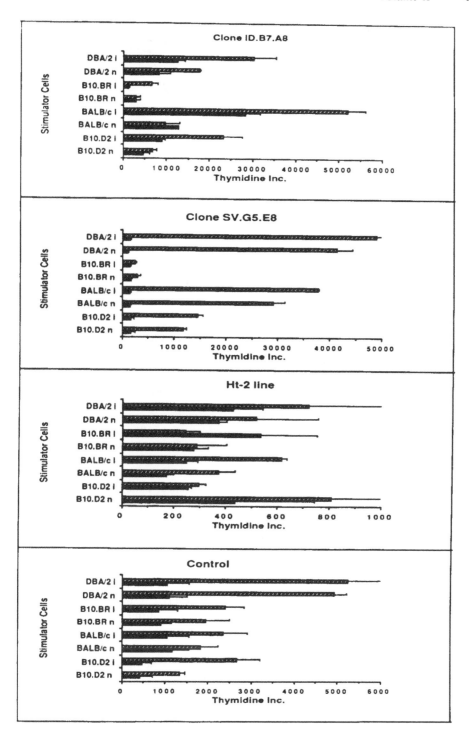

FIGURE 6.   Specificity of 1B4.H6-induced T-cell clones. Clone ID.B7.A8 (T-cell clone which initially had been induced by immunization of B10.D2 mice with 1B4.E6) and clone SV.G5.E8 (a T-cell clone which had been induced by immunization of B10.D2 mice with Sendai virus) were tested for proliferation upon coculture with unmodified (solid bar) or Sendai virus-pretreated (hatched bar) splenocytes derived from naive (n) or Sendai virus-immune (i) mice as described in Figure 5. To control for IL-2 production by the stimulator cell population, the IL-2-dependent HT-2 line (which as was tested in parallel is more sensitive to IL-2 as compared to the two T-cell clones) was tested on the same batch of stimulator cells. To control for background proliferation irradiated stimulator cells (2000 rads) were tested.

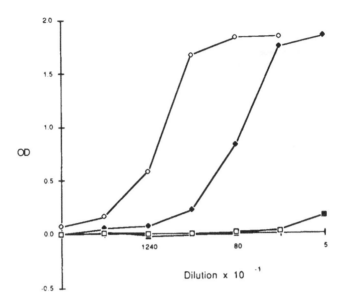

FIGURE 7.   Induction of a B-cell response. Groups of DBA/2 mice were immunized with 10^3 HAU of UV-light inactivated Sendai virus or 200 $\mu\ell$ of 1B4.H6 culture supernatant. These mice, as well as age-matched naive mice, were bled 24 days later. Sera were tested for binding to plates which had been coated with Sendai virus or as a control bovine serum albumine (BSA) in an enzyme-linked immunosorbent assay as described.[57] Normal mouse serum showed no binding to either Sendai virus (OD of 0.02 at a 1:50 dilution) or BSA (OD of 0.009 at a 1:50 dilution). Sendai virus-immune serum tested on Sendai virus (solid circle), tested on BSA (solid square); 1B4.E6-immune serum tested on Sendai virus (open circle), tested for BSA (open square). OD, optical density.

mimicry as defined for B-cell-specific anti-Id is inconceivable. The 1B4.E6/H6 antibody was selected by its binding to a T helper cell clone which obeys H-2 restriction and, thus, by definition, is unable to recognize and bind free antigen (i.e., Sendai virus). The affinity of 1B4.E6/H6 for the T-cell receptor is, thus, higher than the affinity of Sendai virus. In addition, 1B4.E6/H6-induced T cells which differed in their H-2 specificity from Sendai virus-induced T cells (i.e., apparent lack of H-2 restriction). We assume that it is possible to generate anti-idiotypic antibodies to T-cell Id which mimic antigen and, thus, induce a classical H-2-restricted T-cell response. The selection procedures for these anti-Id would have to be different from those chosen for 1B4.E6/H6 (for example, selection by in vitro functions).

2.   **Mimicry of an antigen-H-2 complex:** Assuming that the T-cell receptor has two binding sites to recognize a complex of H-2 + antigen, one might be able to induce an antireceptor antibody with specificity for both binding sites (or at least a proportion of both binding sites). As the 1B4.E6/H6 antibody bound to Sendai virus specific-T cells of different H-2 haplotypes, recognition of an anti-H-2 receptor by 1B4.E6/H6 can be excluded. For the same reason mimicry of an altered H-2 antigen is impossible.

3.   **Recognition of a paratope of a Sendai virus-specific T-cell idiotope:** The most likely explanation is that 1B4.E6 binds to the proportion of the T-cell receptor which recognizes (with low affinity) the foreign antigen. This paratope seems to be commonly used, as nearly 30% of all Sendai virus-specific T-cell clones we tested as well as a significant percentage of bulk culture-derived Sendai virus-specific T cells bound the 1B4.E6/H6 antibody.

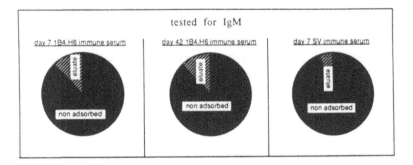

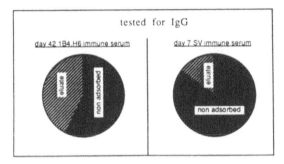

FIGURE 8. Id specificity of the B-cell response. Groups of DBA/2 mice were immunized with 1B4.H6. Sera were harvested 7 or 42 days later. Sera were adsorbed to CNBr-activated Sepharose® beads which had been coupled with 1B4.H6 antibody. The nonadsorbed fraction as well as the adsorbed fraction, which was eluated by addition of 3 *M* potassium thiocyanate (eluate), was tested for antiviral activity in an enzyme-linked immunoadsorbent assay using an antimouse IgM or IgG as second antibody. Day 7 1B4.H6 immune serum had no measurable IgG activity on Sendai virus (data not shown). To adjust for nonspecific binding, serum from Sendai virus-immune DBA/2 mice which, presumably, contains no significant anti-1B4.H6 activity (as confirmed by immunodiffusion) was treated the same way and used as a negative control.

f. Specificity of Anti-Id Induced T cells

The 1B4.E6/H6 antibody presumably induces T cells with different specificities; three types of clones are stimulated:

1. **Classical Sendai virus-specific T cells (which can be induced by Sendai virus as well as 1B4.E6/H6):** These T cells are expected to be H-2 restricted. A percentage of the anti-Id-induced T-cell response was H-2 restricted, as demonstrated by the H-2 preference of the anti-Id-induced delayed-type hypersensitivity response (see Figure 1). The same was shown for cytolytic T cells: when tested at a clonal level, 30 to 40% of all anti-Id-induced T-cell clones lysed the H-2-compatible virus-infected target cell, but failed to lyse allogeneic-infected target cells. This type of T cell might be induced by soluble antibody or upon processing and presentation with H-2 molecules.

2. **Silent Sendai virus-specific T-cell clones (which are normally not induced during an antigen-driven immune response):** T cells which apparently lack H-2 restriction are normally not induced upon immunization with antigen. Lack of H-2 restriction of anti-Id-induced T cells could reflect increased affinity for antigen which would override the requirement for corecognition of H-2. Recognition of the 1B4.E6/H6-expressing hybridoma cell line, presumably, reflects a true lack of H-2 restriction due to high affinity to antigen, in this case the anti-Id. An apparent lack of H-2 restriction due to recognition of H-2 framework determinants seems to be a likely explanation for the

specificity of the proliferative T-cell clones, as those T-cell clones proliferated, though weakly, in response to antigen in association with H-2^d and H-2^b, but not H-2^k. Proliferation could be inhibited by antibodies to Ia framework determinants.

3. **Anti-Id-specific T cells which fail to recognize Sendai virus:** A large percentage of T-cell clones induced with the anti-Id are, presumably, directed to framework or allotype determinants of the 1B4.E6/H6 antibody. Our experiments were designed to select for Sendai virus-specific T cells; we are, thus, not able to estimate the extent of stimulation of "unwanted" clones.

g. Specificity of the Anti-Id-Induced B-Cell Response

Frischknecht et al. showed previously that a T-cell specific auto-antiserum could induce an antigen-specific B-cell response in vivo. His conclusion was that T and B cells carry cross-reactive Id which can serologically be influenced by the same anti-idiotypic reagents. Using a monoclonal anti-T-cell Id, we obtained similar results, i.e., stimulation of an antigen-specific T- and B-cell response in vivo. Our conclusions, however, are somewhat different. Analysis of the antibody response for anti-Id specificity clearly indicated that cross reactivity of T- and B-cell receptors could not explain the induction of an antigen-specific antibody response. Only a marginal percentage of anti-Id-induced IgM antibodies were antigen specific and anti-anti-idiotypic; the majority of the antigen-specific immunoglobulins failed to bind the anti-Id (>80%). We do not fully understand the mechanism of induction of B cells with noncomplementary Id (which has as well been observed upon immunization with B-cell-defined anti-Id[53]). It is unlikely that 1B4.E6/H6-induced B cells were initially anti-anti-idiotypic and upon proliferation and hypermutation lost their anti-Id specificity while retaining antigen specificity. This should be reflected by a loss of anti-Id binding of late IgG antibodies as opposed to early IgM antibodies. Our data show just the opposite. Alternatively, the 1B4.E6/H6 might interrupt a network of idiotypic/anti-idiotypic B and T cells. Release of anti-idiotypic B (or T) cells by binding of 1B4.E6/H6 to idiotypic T cells might trigger an anti-anti-idiotypic B-cell response which is Sendai virus specific, but no longer complementary to 1B4.E6/H6.

2. Listeria monocytogenes

Protective immunity to facultative intracellular bacteria such as *L. monocytogenes* depends on T-cell-mediated immunity. As described above for the Sendai virus system, syngeneic anti-idiotypic antisera with clonotypic activity for an *L. monocytogenes*-specific T helper cell hybridoma were shown to induce protective immunity against a subsequent bacterial infection.[54]

Data obtained with Sendai virus and *L. monocytogenes* clearly show that T-cell-defined anti-idiotypic antibodies can induce protective immunity and, thus, have potential as antiviral vaccines.

V. CANDIDATES FOR ANTI-ID VACCINES

B-cell-defined anti-idiotypic antibodies offer an alternative to classical attenuated or inert vaccines in prevention of infections where protective immunity is mainly dependent on neutralizing antibodies or Class II-restricted T cells. T-cell-defined anti-Id could alternatively be used to prevent infections with viruses or intracellular bacteria whose successful elimination requires cytolytic T cells. Though live attenuated vaccines will undoubtedly continue to be the main source of vaccines, anti-idiotypic antibodies might be an alternative for some infectious agents.

A. Immunization of Nonresponsive Hosts

Infants fail to respond to immunization with bacterial polysaccharide vaccines. In mice this is postulated to be due to an ontogenic delay of a subset of B cells (Lyb-5$^+$ B cells). Though Lyb-5$^-$ B cells with appropriate idiotopes are present at birth, these B cells seem to be tolerized by environmental polysaccharides. Administration of anti-idiotypic antibodies has been shown to break this stage of tolerance and, thus, increase resistance to bacterial infections such as with *Escherichia coli.*

B. Immunization to Quantitatively Limited Antigens

Conventional vaccines cannot be produced to parasites which cannot be grown in sufficient quantities in vitro. Even though methods of modern molecular biology might provide genetically engineered peptides, differences in glycosylation or tertiary structure might influence their immunogenicity and, thus, forestall their use as vaccines. Monoclonal anti-idiotypic antibodies, which can be obtained in unlimited quantities, would be an alternative in the vaccination against these antigens.

C. Immunization to Tumorigenic Viruses

Some viruses such as herpesviruses or human T lymphotropic viruses (HTLV), which can integrate into the host genome and cause malignancies, might be too hazardous to allow use in humans even after attenuation or inactivation. Anti-idiotypic antibodies either to T or B cells might be used to induce protective immunity without undue risk.

D. Immunization to Immunosuppressive Reagents

Proteins associated with the envelopes of retroviruses such as HTLV III or feline leukemia virus have been shown to cause immunosuppression.[55] Thus, even highly purified proteins of these viruses which (due to lack of genomic material) will not carry the risk of tumor induction might be unsuitable as vaccines.

E. Immunization to Immunoevasive Reagents

Some viruses such as influenza A virus have been shown to undergo rapid mutations and, thus, evade the immune system. As the mutations affect mainly envelope glycoproteins such as the hemagglutinin molecule, which is the target for neutralizing antibodies, B-cell-mediated immunity such as acquired by vaccination with inactivated vaccines or subunit vaccines often fails to provide adequate protection. T cells in contrast to B cells are primarily directed to internal virus proteins (such as the nucleoprotein) which are highly conserved. Cytolytic T cells have, thus, been shown to transfer cross-reactive protection to all influenza A virus strains.[56] T- (or B-) cell-defined anti-idiotypic antibodies to a cross-reactive epitope might be used to induce protection to variants of mutating viruses.

VI. OPEN QUESTIONS

Though T- and B-cell-defined anti-idiotypic antibodies have been shown to induce protective immunity in laboratory animals, many questions remain to be solved.

1. The use of xenogeneic antibodies will evoke allergic reactions. Construction of chimeric antibodies might overcome this problem.
2. Most investigations have concentrated on the ability of anti-Id to induce an acute response. For practical applications vaccines must be able to induce long-lasting immunity. The duration of the anti-Id-induced immunity should, thus, be further investigated.
3. Though anti-Id of the IgG and IgM class have been used successfully to induce immunity, the role of the isotype of an anti-idiotype reagent should further be investigated.
4. Most groups which have used B-cell-defined anti-idiotypic reagents tested for the

induction for neutralizing antibodies, but not for T-cell-mediated responses. Only one group tested for stimulation of cytolytic T lymphocytes and obtained negative results. It would be important to know if B-cell-defined anti-idiotypic antibodies are, in general, unable to induce Class I-restricted T cells.

5.   Limited data are available on the use of T-cell-defined anti-idiotypic antibodies. Further investigations are needed to classify T-cell idiotopes.

6.   The potential use of anti-Id in suppression of unwanted immune response should be investigated.

All these questions can be easily answered with currently available methodology. Such investigations are currently underway and will allow for evaluation of those reagents as vaccines.

REFERENCES

1. **Jerne, N. K.**, *Idiotypes — Antigens on the Inside*, Westen-Schrum, I., Ed., Eolt. Rode, Basel, 1981, 12.
2. **Cerny, J. and Kelsoe, G.**, Priority of the anti-idiotypic responses after antigen administration: after fact or intriguing network mechanism?, *Immunol. Today*, 5, 259, 1984.
3. **Urbain, J., Wikler, M., Franssen, J. D., and Collignon, C.**, Idiotypic regulation of the immune system by the induction of antibodies against anti-idiotypic antibodies, *Proc. Natl. Acad. Sci. U.S.A.*, 74, 5126, 1977.
4. **Cazenave, P.-A.**, Idiotypic-anti-idiotypic regulation of antibody synthesis in rabbits, *Proc. Natl. Acad. Sci. U.S.A.*, 74, 4122, 1977.
5. **Jerne, N. K.**, Towards a network theory of the immune system, *Ann. Immunol.*, 125C, 373, 1974.
6. **Zinkernagel, R. M. and Doherty, P. C.**, Restriction of in vitro T cell-mediated cytotoxicity in lymphocytic choriomeningitis within a syngeneic or semi-allogeneic system, *Nature (London)*, 251, 547, 1974.
7. **Doherty, P. C. Blanden, R. V., and Zinkernagel, R. M.**, Specificity of virus immune effector T cells for H-2K or H-2D compatible interactions. Implications for H-antigen diversity, *Transplant. Rev.*, 29, 89, 1976.
8. **Hood, L., Kronenberg, M., and Hunkapiller, T.**, T cell antigen receptor and the immunoglobulin supergene family, *Cell*, 40, 225, 1985.
9. **Wu, T. T. and Kabat, E. A.**, An analysis of the sequences of the variable regions of Bence-Jones protein and myeloma light chains and their implications for antibody complementary, *J. Exp. Med.*, 132, 211, 1970.
10. **Segal, D. M., Padlan, E. A., Cohen, G. H., Rudikoff, S., Potter, M., and Davies, D. R.**, The three-dimensional structure of a phosphocholine-binding mouse immunoglobulin Fab and the nature of the antigen binding site, *Proc. Natl. Acad. Sci. U.S.A.*, 71, 4298, 1974.
11. **Yanagi, Y., Yoshikai, Y., Leggett, K., Clark, S. P., Aleksander, I., and Mak, T. W.**, As human T cell cDNA clone encodes a protein having extensive homology to immunoglobulin chains, *Nature*, 308, 145, 1984.
12. **Hedrick, S. M., Cohen, D. I., Nielson, E. A., and Davies, M. M.**, Isolation of cDNA clones encoding T cell specific membrane associated proteins, *Nature*, 308, 149, 1984.
13. **Ashwell, J. D. and Schwartz, R. H.**, T cell recognition of antigen and the Ia molecule as a ternary complex, *Nature*, 320, 176, 1986.
14. **Watts, T. H., Gaub, H. E., and McConnel, H. M.**, T cell-mediated association of peptide antigen and major histocompatibility complex protein detected by energy transfer in an evanescent wave-field, *Nature*, 320, 179, 1986.
15. **Coszena, H., Julius, M. H., and Augustin, A. A.**, Idiotypes as variable region markers: analogies between receptors on PC specific T and B lymphocytes, *Immunol. Rev.*, 34, 3, 1977.
16. **Eichmann, K. and Rajewsky, K.**, Induction of T and B cell immunity by anti-idiotypic antibody, *Eur. J. Immunol.*, 5, 66, 1975.
17. **Eichmann, K., Ben-Neriah, Y., Hetzelberger, D., Polke, C., Givol, D., and Lonai, P.**, Correlated expression of Vh framework and Vh idiotypic determinants on T helper cells and on functionally undefined T cells binding group A streptococcal carbohydrate, *Eur. J. Immunol.*, 10, 105, 1980.
18. **Haskins, A., Hanmun, C., White, J., Roehnn, U., Kubo, R., Kappler, J., and Marrack, P.**, The antigen-specific major histocompatibility complex restricted receptor on T cells. I. Isolation with a monoclonal antibody, *J. Exp. Med.*, 157, 1149, 1984.

19. **Thomas, W. R., Morahan, G., Walker, J. D., and Miller, J. F. A. P.,** Induction of delayed-type hypersensitivity to azobenzenearsonate by a monoclonal anti-idiotype antibody, *J. Exp. Med.*, 153, 743, 1981.

20. **Sharpe, A. H., Gaulton, G. N., McDade, K. K., Fields, B. B., and Green, M. I.,** Syngeneic monoclonal antiidiotype can induce cellular immunity to reovirus, *J. Exp. Med.*, 160, 1195, 1984.

21. **Kaye, J., Porcelli, S., Tite, J., Jones, B., and Janeway, C. A.,** Both a monoclonal antibody and a serum specific for determinants unique to individual cloned helper T cell live can substitute for antigen and antigen-presence cells in the activation of T cells, *J. Exp. Med.*, 158, 836, 1983.

22. **Ertl, H. C. J. and Finberg, R. W.,** Sendai virus specific T cell clones: induction of cytolytic T cells by anti-idiotypic antibody directed against a helper T cell clone, *Proc. Natl. Acad. Sci. U.S.A.*, 81, 2850, 1984.

23. **Saito, J. and Rajewsky, K.,** Helper T cells reacting to idiotype on IgG but not IgM, *J. Exp. Med.*, 162, 1399, 1985.

24. **Sy, M.-S., Bach, B. A., Dohi, Y., Nisonoff, A., Benacerraf, B., and Greene, M. I.,** Antigen and receptor driven regulatory mechanisms. I. Induction of suppressor T cells by anti-idiotypic antibodies, *J. Exp. Med.*, 150, 1216, 1979.

25. **Weinberger, J. L., Germain, R. V., Benacerraf, B., and Dorf, M. E.,** Hapten specific T cell responses to 4-hydroxy-3-nitrophenyl acetyl. V. Role of idiotypes in the suppressor pathway, *J. Exp. Med.*, 152, 161, 1980.

26. **Neuberger, M. S., Williams, G. T., Mitchell, E. B., Jouhal, S. S., Flanagan, J. G., and Rabbitts, T. H.,** A hapten specific chimaeric IgE antibody with human physiological effector function, *Nature*, 314, 268, 1985.

27. **Schecter, Y., Mason, R., Elias, D., and Cohen, J. R.,** Auto-antibodies to insulin receptor spontaneously develop as anti-idiotypes in mice immunized with insulin, *Science*, 216, 242, 1982.

28. **Co, M. S., Gaulton, G. N., Tominaga, A., Homay, C. J., Fields, B. N., and Greene, M. I.,** Structural similarities between the mammalian β-andrenergic and reovirus type 3 receptors, *Proc. Natl. Acad. Sci. U.S.A.*, 82, 5315, 1985.

29. **Sacks, D. L., Esser, K. M., and Sher, A.,** Immunization of mice against African trypanosomiasis using antiidiotypic antibodies, *J. Exp. Med.*, 155, 1108, 1982.

30. **Sacks, D. L. and Sher, A.,** Evidence that anti-idiotype induced immunity to experimental African trypomasomiasis is genetically restricted and requires recognition of combining site related idiotypes, *J. Immunol.*, 131, 1511, 1983.

31. **Grzych, M. M., Capron, M., Lambert, P. H., Dissous, C., Torres, S., and Capron, A.,** An anti-idiotype vaccine against experimental schistosomiasis, *Nature*, 316, 74, 1985.

32. **Kennedy, R. C., Adler-Stortz, K., Henkel, R. D., Sanchez, Y., Melnick, J. L., and Dreesman, G. R.,** Immune response to hepatitis B surface antigen: enhancement by prior injection of anti-idiotype antibodies, *Science*, 221, 853, 1983.

33. **Reagan, K. J., Wunner, W. H., Wiktor, T. J., and Koprowski, H.,** Anti-idiotypic antibodies induce neutralizing antibodies to rabies virus glycoprotein, *J. Virol.*, 48, 660, 1983.

34. **UytdeHaag, F. G. C. M. and Osterhaus, A. D. M. E.,** Induction of neutralizing antibody in mice against poliovirus type II with monoclonal anti-idiotypic antibody, *J. Immunol.*, 134, 1225, 1985.

35. **Roehring, J. T., Hunt, A. R., and Mathews, J. H.,** Identification of anti-idiotype antibodies that mimic the neutralization site of Venezuelan equine encephalomyelitis, Abstract, High Technology Route to Virus Vaccines, Houston, Tex., 1984, 32.

36. **Sutton, L. S. and Brooke, C. C.,** Venezuelan equine encephalomyeletis due to vaccination in man, *J. Am. Med. Assoc*, 155, 1473, 1954.

37. **Francotte, M. and Urbain, J.,** Induction of anti-tobacco mosaic virus antibodies in mice by rabbit anti-idiotypic antibodies, *J. Exp. Med.*, 160, 1485, 1984.

38. **Ertl, H. C. J., Greene, M. I., Noseworthy, J. H., Fields, B. N., Nepom, T. J., Spriggs, D. R., and Finberg, R. W.,** Identification of idiotypic receptors on reovirus specific cytolytic T cells, *Proc. Natl. Acad. Sci. U.S.A.*, 79, 7479, 1982.

39. **Sharpe, A. J., Gaulton, G. N., Ertl, H. C. J., Finberg, R. W., McDade, K. K., Fields, B. N., and Greene, M. I.,** Cell receptor for the mammalian reovirus. Reovirus specific cytolytic T cell lines that have idiotypic receptors recognize anti-idiotypic B cell hybridomas, *J. Immunol.*, 134, 2702, 1985.

40. **McNamara, M. K., Ward, R. E., and Koehler, H.,** Monoclonal idiotope vaccine against Streptococcus pneumonia infection, *Science*, 226, 1325, 1984.

41. **Stein, K. and Soderstrom, T.,** Neonatal administration of idiotype or anti-idiotype primes for protection against Escherichia coli K13 infection in mice, *J. Exp. Med.*, 160, 1001, 1984.

42. **Gorczynski, R. M., Kennedy, M., Polidoulis, I., and Price, G. B.,** Altered tumor growth *in vivo* after immunization of mice anti-tumor-antibodies, *Cancer Res.*, 44, 3291, 1984.

43. **Nepom, G. J., Nelson, K. A., Holbeck, S. L., Hellström, J., and Hellström, V. I.,** Induction of immunity to a human tumor marker by *in vivo* administration of anti-idiotypic antibodies in mice, *Proc. Natl. Acad. Sci. U.S.A.*, 81, 2864, 1984.

44. **Kennedy, R. C., Dreesman, G. R., Butel, J. S., and Lanford, R. E.,** Suppression of *in vivo* tumor formation induced by Simian virus 40-transformed cells in mice receiving anti-idiotypic antibodies, *J. Exp. Med.*, 161, 1431, 1985.

45. **Kennedy, R. C., Alder-Stortz, K., Burns, J. W., Henkel, R. D., and Dreesman, G. R.,** Anti-idiotype modulation of herpes simplex virus infection leading to increased pathogenicity, *J. Virol.*, 50, 951, 1984.

46. **Ada, G. L., Leung, K.-N., and Ertl, H. C. J.,** An analysis of effector T cell generation and function in mice exposed to influenza A or Sendai virus, *Immunol. Rev.*, 58, 6, 1981.

47. **Bona, C. and Moran, J.,** Idiotype vaccines, *Ann. Inst. Pasteur/Immunol.*, 136C, 21, 1985.

48. **Takemori, J. and Rajewskey, K.,** Mechanism of neonatally induced idiotype suppression and its relevance for the acquisition of self-tolerance, *Immunol. Rev.*, 179, 104, 1984.

49. **Frischknecht, H., Binz, H., and Wigzell, H.,** Induction of specific transplantation immune reaction using anti-idiotypic antibodies, *J. Exp. Med.*, 147, 500, 1978.

50. **Ryser, J.-E., Cerottini, J.-C., and Brunner, K. T.,** Generation of cytolytic T lymphocytes *in vitro*. Induction of secondary CTL responses in primary long-term MLC by supernatants from secondary MLC, *J. Immunol.*, 120, 370, 1978.

51. **MacDonald, H. R., Cerottini, J. C., Ryser, J.-E., Maryanski, J. L., Taswell, C., Widmer, M. B., and Brunner, K. T.,** Quantitation and cloning of cytolytic T lymphocytes and their precursors, *Immunol. Rev.*, 51, 93, 1980.

52. **Ertl, H. C. J. and Finberg, R. W.,** Characteristics and functions of Sendai virus specific T cell clones, *J. Virol.*, 50, 425, 1984.

53. **Ertl, H. C. J., Homans, E., Tournas, S., and Finberg, R. W.,** Sendai virus specific T cell clones. V. Induction of virus specific response by antiidiotypic antibodies directed against a T helper cell clone, *J. Exp. Med.*, 159, 1778, 1984.

54. **Kaufman, S. H. E., Eichman, K., Müller, J., and Wrazel, L. J.,** Vaccination against the intracellular bacterium Listeria monocytogenes with a clonotypic anti-serum, *J. Immunol.*, 134, 4123, 1985.

55. **Laurence, J.,** The immune system in AIDS, *Sci. Am.*, 253, 84, 1985.

56. **Yap, K. L., Ada, G. L., and McKenzie, I. F. C.,** Transfer of specific cytotoxic T lymphocytes protects mice inoculated with influenza virus, *Nature (London)*, 273, 238, 1978.

57. **Ertl, H., Gerlich, W., and Koszinowski, U.,** Detection of antibodies for Sendai virus by enzyme linked immunosorbent assay, *J. Immunol. Methods*, 28, 163, 1976.

Chapter 4

IDIOTYPE CONNECTIVITY OF ANTIBODY RESPONSES SPECIFIC FOR SELF- AND NONSELF ANTIGENS

D. S. Dwyer

TABLE OF CONTENTS

I. INTRODUCTION

The diverse elements of the immune system, particularly the specific antigen recognition structures borne by B and T cells, must be highly organized so that efficient interlymphocyte communication and regulation can occur. Jerne described this organization with a unifying concept known as the idiotype. (Id) network.[1] In this model, the variable regions of antigen binding molecules are organized into networks of complementary and interacting structures. The Id network theory has had tremendous heuristic value and numerous studies have now confirmed that anti-idiotypes (anti-Id) do participate in the regulation of specific immune responses.[2-5]

Implicit in the network theory is the concept that antibodies against disparate antigens are idiotypically connected. Indeed, several groups have reported that distinct antibodies against different antigens or against different epitopes on the same antigen share Id,[6-8] confirming the above notion of interconnectivity among antibodies. The idiotypic connectivity among antibodies must extend to autoantibodies as well, thereby linking autoimmune responses to self-antigens with immune responses to common nonself-antigens or pathogens. This topic will be discussed in detail in this chapter along with the implications for the initiation of autoimmunity.

The autoimmune disease studied in this report is myasthenia gravis (MG). Autoantibodies against the acetylcholine receptor (AChR) produce the impairment in neuromuscular transmission observed in patients with this disease.[9,10] These autoantibodies can be detected in the serum of 90% of myasthenic patients.[11,12] However, the cause of the autoimmune response is not known. Some patients with MG produce specific anti-idiotypic antibodies in addition to autoantibodies against the AChR.[13] These anti-Id react specifically with a subset of anti-AChR antibodies and often bear a relationship to key indicators of disease activity such as the anti-AChR antibody titer or the clinical symptoms.[14] The anti-Id are thought to contribute to the regulation of the anti-AChR response. Therefore, this human autoimmune disease provides an opportunity to study Id networks in relation to the production of autoantibodies.

This chapter will discuss the idiotypic connections between self- and nonself antigens. Based on this principle, we will propose a model whereby autoimmunity is initiated by the normal operation of Id networks. It will be shown that an anti-Id in one antigen system is a potentially harmful autoantibody in another. Finally, the concept of super organizer antibodies and the special case of epibodies will be developed to describe the organization and functional hierarchy in Id networks.

II. IDIOTYPIC CONNECTIVITY: THE AChR-DEXTRAN NETWORK

Recently, we have reported that the immune responses to the AChR and the polysaccharide $\alpha 1,3$ dextran (DEX) are connected via extensive Id-anti-Id interactions.[15,16] The AChR-DEX network described in these reports consists of 15 monoclonal antibodies (Mabs) which form an elaborate web of interacting Id. Based on this network, it was predicted that patients with MG might have anti-DEX antibodies in their serum and this suggestion was confirmed.[16] Approximately 15% of the patients with MG have serum antibodies against DEX, whereas only 1 of 75 control samples has been positive. Because the DEX epitope is associated with certain common enteric bacteria,[17] we suggested that these bacteria are involved in the initiation of MG.

A simplified version of the AChR-DEX network is shown in Figure 1 to illustrate how the bacterial epitopes (DEX) could initiate the anti-AChR response. A key antibody in this scheme is SR11 which was originally described as a human monoclonal anti-Id against ACR24, a BALB/c Mab specific for the AChR.[13] Elsewhere, we have demonstrated that there are idiotypic counterparts to ACR24 in sera from myasthenic patients.[13] Thus, this

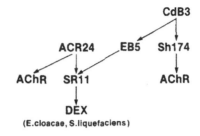

FIGURE 1. The AChR-DEX network: a model for the initiation of MG. A portion of the AChR-DEX network is depicted here; for more detail see References 15 and 16. The arrows represent significant binding as measured in the ELISA. All possible combinations of binding were tested and only those shown here were significant; greater than four standard deviations above the BSA background. ACR24 (γ1, κ) and Sh174 (γ2b, κ) are BALB/c Mabs against the AChR. EB5 (γ1, κ) is a BALB/c monoclonal anti-Id specific for SR11. SR11 (μ,κ) is a human monoclonal anti-Id initially selected because it bound to ACR24. CdB3 (μ,κ) is a syngeneic anti-Id raised against Sh174.

murine Id is expressed in patients with MG and may be important in the development of MG. Recently, it was discovered that SR11 reacts specifically with DEX, in addition to ACR24, and that the binding to these two antigens is accomplished by separate recognition sites.[15] In fact, SR11 binding to DEX is inhibited by the trisaccharide, nigerotriose, which is the smallest carbohydrate unit bearing the DEX epitope indicating the exquisite specificity of this antibody. In Figure 1, ACR24 is portrayed as an anti-Id against SR11 because this interaction is bidirectional. Based on this scheme, we propose that the autoimmune response to the AChR is initiated in the following manner. DEX epitopes on certain bacteria would stimulate the production of antibodies like SR11. To regulate the anti-DEX response, anti-Id like ACR24 and EB5 would be produced. However, ACR24, in addition to being an anti-Id, is an autoantibody against the AChR. The production of ACR24 would be the initial step in autoantibody formation and because ACR24 is an anti-Id, its initial production might not be subject to suppression by AChR-specific suppressor T cells. The other anti-Id, EB5, is idiotypically linked (via CdB3) to the anti-AChR antibody Sh174. Therefore, the autoimmune response to the AChR would be amplified if, via network perturbations, antibodies like Sh174 were produced. This initial wave of autoantibodies, together with other components such as complement or macrophages, would damage the muscle fiber membrane releasing the AChR or proteolytic fragments and stimulating the production of more autoantibody. Plotz had independently proposed a similar mechanism for the initiation of autoimmunity[18] whereby an anti-Id in one system is an autoantibody in another. We have presented evidence that this may actually occur.

III. IDIOTYPIC CONNECTIONS BETWEEN ANTIBODIES AGAINST ENVIRONMENTAL ANTIGENS AND AUTOANTIGENS

One question raised by these findings concerns the specificity and functional relevance of the network connections described here. It has previously been reported that the antibody binding outlined in Figure 1 is highly specific and idiotypic in nature.[16] Within a given network, most of the antibodies do not extensively interact with other members of the same network (e.g., Sh174), although they may interact with members of other antigen networks. In contrast, a few of the antibodies within a network do interact with a number of other members of the same network (e.g., CdB3) and also interact with other Id networks. These multispecific antibodies impose a hierarchy on the organization of the Id network. We have referred to these antibodies as super organizers, because they both organize the antibodies within a given network and they link antibodies against disparate antigens. These antibodies

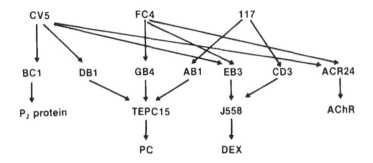

FIGURE 2.   Idiotypic connectivity between self- and nonself antigens. Significant antibody interactions are again represented with an arrow. Binding was assessed with an ELISA where antigen was coated onto polyvinyl plates. BC1 (γ 2a, κ) is a BALB/c Mab specific for P$_2$ protein, whereas TEPC15 (α, κ) and J558 (α, λ) are BALB/c myeloma proteins specific for PC and DEX, respectively. DB1 (γ 2a, κ), GB4 (γ1, κ), and AB1 (α 1, κ) are anti-Id derived from A/J mice and react with the T15 Id. EB3 (γ1, κ) is an A/J anti-Id specific for J558, whereas CD3 (A/J γ1, λ) recognizes an Id shared by anti-DEX antibodies. CV5 (μ,κ) is a syngeneic anti-Id raised against BC1. FC4 (μ,λ) is derived from the spleens of 3-day-old BALB/c mice. Finally, 117 (μ,κ) was obtained from adult BALB/c mice immunized with DEX based on its interaction with CD3.

appear to have other special properties which will be discussed more fully in the final section. An example of several super organizers and their relationship to various antigen systems is shown in Figure 2. This scheme illustrates two important points. First, the super organizers impose an organizational hierarchy on the network, as mentioned before, by virtue of their multispecificity. Second, these anti-Id connect the immune responses to diverse antigens, including autoantigens. The Id connections between antibodies against the AChR and DEX have been discussed earlier. It can be seen that the anti-AChR response is also idiotypically related to the phosphorylcholine (PC) antigen system (via FC4). Furthermore, these data demonstrate the generality of this phenomenon. As shown here, CV5 connects the immune response to the P$_2$ protein of peripheral myelin with the response to PC, DEX, and even the AChR. An autoimmune response to the P$_2$ protein may contribute to the pathogenesis of the Guillain-Barre syndrome.[19,20] In summary, antibodies against two autoantigens have been idiotypically linked to antibodies against common bacterial epitopes.

Others have reached a similar conclusion using a different paradigm. Avrameas et al. have studied the specificity of natural antibodies, especially those which bind to self-antigens.[21,22] Some of the antibodies isolated from unimmunized neonatal and adult mice are broadly reactive and a subset of these antibodies reacts with self-antigens such as thyroglobulin, tubulin, and myosin. Recently, Lymberi et al.[23] determined that there is a high incidence of cross-reactive Id among these natural autoantibodies. These studies provide another example of shared idiotypy between antibodies against self- and nonself-antigens. If these Id connections represent functional connections as well, then Id-mediated regulation becomes a more complex process because it simultaneously impinges on numerous antigen systems including self-antigens. Imprecise regulation or a deficiency in the other contributing regulatory elements (such as T suppressor cells) could lead to the expression of autoantibodies as discussed earlier.

IV. EPIBODIES: A SPECIAL TYPE OF SUPER ORGANIZER ANTIBODY

During these studies of Id connections, we discovered that a subset of the super organizer antibodies exhibited unique binding characteristics. These antibodies recognize a cross-reactive Id found on certain anti-AChR Mabs and, in addition, they bind to the AChR. Therefore, these antibodies bind to both Id and to the antigen that the Id-bearing antibodies recognize. Antibodies with these properties have been named epibodies by Bona et al.[24] The

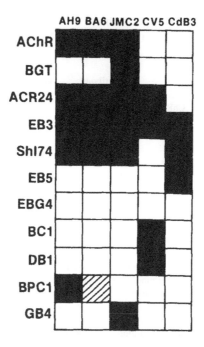

FIGURE 3. Summary of the binding patterns of five super organizer antibodies. All of these antibodies have been described elsewhere in this paper except EBG4 (γ 2a, κ), which is a BALB/c Mab directed against the AChR. The darkened squares represent significant binding measured in the ELISA. The cross-hatchings represent weak but reproducible binding.

original description of epibodies was based on the observation that certain anti-Id reacting with a shared Id on human rheumatoid factors also bound to the Fc portion of IgG. These findings were confirmed by Chen et al.[25] who showed that there is shared sequence homology between the cross-reactive Id and the Fc portion of IgG. However, it may not always be the case that the epibody binds to the same determinant shared by the Id and the antigen, as will be shown below. One example has been provided by Kang and Kohler[26] who discovered an epibody which binds to both PC and to anti-PC antibodies. Interestingly, this antibody bears the T15 Id characteristic of BALB/c anti-PC antibodies and binds to itself. This unusual property led Kang and Kohler to coin the term "autobody" — a special self-binding epibody.

The epibodies described thus far do not bind to autoantigens other than immunoglobulin epitopes. However, we have found several epibodies which bind to the AChR, in addition to binding to anti-AChR antibodies, indicating that these epibodies are true autoantibodies. Figure 3 summarizes the specificity pattern of three epibodies in comparison to two other super organizers. These three epibodies were derived in completely different ways, and yet their binding patterns are fairly similar. AH9 was obtained from mice used in an experiment to study the effects of anti-Id treatment on the immune response to the AChR. The administration of anti-Id against AChR-specific antibodies prior to immunization with AChR reduces the subsequent immune response to the AChR.[27] The effectiveness of this treatment appears to be related to the induction of endogenous anti-Id.[27] AH9 is one example of the induced anti-Id obtained from pretreated mice. After multiple limiting dilution steps to insure the clonality of AH9, this antibody retained its original dual specificity for the AChR and for the anti-AChR antibodies, ACR24 and Sh174. In addition, AH9 binds to EB3 and to BPC1, a Mab of unknown specificity, but does not bind to other members of a large panel of antibodies with similar and different isotypes.

Table 1
INHIBITION OF AH9 BINDING

	Antibody	Inhibition of binding (%)
AH9 to AChR	ACR24	1
	EB3	− 4
	BPC1	63
AH9 to ACR24	ACR24	0
	EB3	2
	BPC1	0

Note: Several antibodies were tested for their ability to inhibit binding of AH9 to either the AChR or to ACR24. The upper set of data was obtained with an immunoprecipitation assay which has been described elsewhere.[42] Inhibition of binding to ACR24 was determined in an ELISA. Various dilutions of inhibiting antibody were tested in these studies. The data are expressed as the percentage inhibition of binding as compared to a control standard at a final concentration of 50 μg/mℓ of inhibitor.

It appears that AH9 recognizes these diverse molecules via two separate binding sites. These data are presented in Table 1. Binding of AH9 to AChR is inhibited by BPC1, but not by ACR24 or EB3. Furthermore, AH9, recognizes epitopes on soluble AChR and BPC1, whereas ACR24 and EB3 are bound only in the solid phase assay as witnessed by their inability to inhibit binding to ACR24 coated onto the polyvinyl plate. This special property of many super organizer antibodies will be elaborated on in the final section. In summary, AH9 is an epibody which reacts with a true autoantigen (the AChR) and with several Id-bearing antibodies. Unlike the epibodies described by Bona et al.[24] and Chen et al.,[25] AH9 does not simply recognize a shared amino acid sequence epitope present on each of the target molecules.

As seen in Figure 3, the specificity pattern of BA6 is very similar to that of AH9. BA6 was derived from the liver of a 2-day-old BALB/c mouse. It was selected on the basis of its interaction with EB3 and was later found to react with Sh174 and ACR24 as well. Because it was similar to AH9, BA6 was then tested for binding to the AChR. BA6 binds strongly to muscle AChR (see below), but very weakly to BPC1.

JMC2 is the last epibody depicted in Figure 3. This Mab was raised by immunizing mice with α-bungarotoxin (BGT). During screening of the fusion, one antibody bound to both BGT and to the AChR. This hybridoma was cloned and recloned twice and the final product was JMC2. Throughout cloning, all subclones retained the capacity to bind both BGT and AChR. Upon further testing, it was discovered that JMC2 also bound to ACR24, EB3, Sh174, and GB4, but not to other members of our panel of Mabs. JMC2 is an epibody because it binds to AChR and to an Id present on anti-AChR antibodies (ACR24 and Sh174). In addition, JMC2 binds to another pair of complementary molecules, the AChR and BGT. BGT is a specific, high affinity ligand for the AChR. Thus far, epibodies have been defined as anti-Id which also bind to the original antigen. The binding of JMC2 to BGT and the AChR is the first example of an antibody binding to a pair of complementary molecules where one of the pair is not an antibody. Therefore, an epibody may be more broadly defined as an antibody which reacts with both members of a pair of complementary structures, regardless of whether the pair is composed of antibody-antigen or ligand-receptor molecules.

Table 2
IMMUNOPRECIPITATION OF
ELECTROPLAX AND MUSCLE AChR

	Electroplax (cpm)	Muscle (cpm)
CON	240	412
BA6	343	1010
AH9	426	1002
AH8	1313	614

Note: Antibodies were incubated with radiolabeled AChR and immunoprecipitated according to described procedures.[42] AChR purified from *Narcine* electroplax or a crude detergent extract of denervated rat muscle was used in these studies. A standard concentration of 0.04 pmol of AChR was added to the assays. The binding data for AH8, another anti-AChR Mab, are shown for comparison. The background value for the electroplax AChR is lower, because purified AChR was used.

The specificity patterns for two additional super organizers are shown in Figure 3 for comparison with the epibodies. CV5 was raised by immunizing mice with BC1, a Mab directed against the P_2 protein of myelin. CV5 cross reacts with ACR24, EB3, and DB1. CdB3 was obtained from mice immunized with the anti-AChR Mab, Sh174. In addition, CdB3 binds to EB3, EB5, and AC3 (not shown here). Neither of these antibodies bind to the AChR, although they are similar to the epibodies shown here. Therefore, the epibodies can be considered a subset of super organizer antibodies which bridge the immune responses to self- and nonself-antigens.

A. Similarities between Neonatally and Adult-Derived Epibodies

As mentioned earlier, the epibodies AH9 and BA6 have very similar properties. Both are BALB/c IgM, κ Mabs which bind to the AChR, ACR24, EB3, and Sh174. AH9 binds strongly to BPC1, whereas BA6 reacts very weakly with this Mab. In addition, these two antibodies are very similar in their recognition of fish vs. muscle AChR. Although AChR from electric organs of rays and that from mammalian skeletal muscle are 90% homologous,[28,29] there are large antigenic differences between these two proteins.[30] Data in Table 2 illustrate this point and show the further similarities between AH9 and BA6. Both antibodies immunoprecipitate AChR from electroplax and muscle, however, immunoprecipitation is far greater with muscle than with electroplax-derived receptor. Therefore, both of these antibodies show the same preference for muscle AChR.

The similarities between these two antibodies raise some important issues. BA6 was derived from the neonatal B-cell repertoire, whereas AH9 was obtained from an adult mouse. These two antibodies bind to muscle AChR and are, therefore, autoantibodies. According to the clonal abortion or clonal anergy theory,[31] autoreactive B cells are eliminated during ontogeny to prevent the development of autoimmunity later in life. The similarities between AH9 and BA6 would argue strongly against this notion. The binding properties of AH9 have not altered appreciably from those of the neonatally derived BA6 antibody. Even if these antibodies are not identical (as shown in the next section), their similarities are striking enough to consider that AH9 and BA6 are derived from similar or identical germline genes. Yet, even though antibodies such as BA6 originally express autoreactivity, very similar antibodies (AH9) are present in adult lymph nodes, suggesting that BA6 is not eliminated

Table 3
AChR SUBUNIT SPECIFICITY OF
AH9 AND BA6

	40K	50K	62K	AChR
AH9	0.351	0.002	0.004	0.550
BA6	0.019	0.234	0.130	0.208

Note: The subunits were isolated from an electro-phoresis gel after denaturation and reduction of the AChR and have been referred to by their apparent molecular weight. In this particular case, the β and γ chains ran very close together and could not be separated during processing of the gel. Therefore, they have been pooled for this analysis (50K). Subunits were coated onto ELISA plates and specific antibody binding was measured. The OD_{405} values after subtraction of a BSA background are presented here.

from the repertoire. In view of the extensive Id connectivity between self- and nonself-antigens described here, deletion of such autoreactive clones would be impossible. Thus, an autoantibody in one antigen system may be an anti-Id in another antigen system and may be necessary for idiotypic regulation in this and other systems. *A priori,* it would be impossible to determine which autoantibodies are crucial for idiotypic regulation and which are not. Probably the only B cells that are truly refractory to self-antigens are those which react with antigens, such as serum albumin, that continuously circulate in body fluids in large quantities. Presumably, even this tolerance can be breached under the right circumstances rendering the idea of clonal abortion of autoreactive B cells untenable. Self-tolerance is more likely maintained through an active, dynamic mechanism including T suppressor and helper activity.

B. Lack of Complete Identity between AH9 and BA6

Further studies were undertaken to determine whether AH9 and BA6 are, indeed, identical. Because both of these antibodies bind to the AChR and the AChR is composed of four different subunits, the fine specificity of these antibodies could be examined. Antibody binding to each of the individual subunits was measured in an ELISA. These results are shown in Table 3. AH9 binds exclusively to the α subunit of the receptor, whereas BA6 reacts most strongly with the β, γ, and δ subunits. Thus, it is clear that the fine specificity differs for these two epibodies.

One final test of identity between these antibodies relied on their reactivity with two anti-Id raised against BA6. From mice immunized with BA6, two specific anti-Id were obtained, TW3 and TW9. These antibodies were tested for binding to purified AH9 and the results are shown in Table 4. TW3 and TW9 react strongly with BA6, but do not bind to AH9. Therefore, it can be concluded that BA6 and AH9 are not identical, although they have very similar specificity profiles. Ultimately, studies of their sequences will determine how these two antibodies are related and whether they are derived from common germline genes.

V. SUPER ORGANIZER ANTIBODIES: PERSPECTIVES

A. Unique Features of These Antibodies

One purpose of this chapter has been to describe the idiotypic connections between self-

Table 4
IDIOTYPIC
DIFFERENTIATION
OF AH9 AND BA6

	AH9	BA6
TW 3.2	0.064	1.001
TW 9.7	0.023	0.821

Note: Specific monoclonal anti-Id were produced against BA6. These antibodies were tested for binding to AH9 in the ELISA and OD_{405} values are presented here.

and nonself-antigens and to present one model whereby this connectivity could lead to the initiation of autoimmune disease. In large part, the idiotypic connectivity is established by a class of anti-Id which we have called super organizers. These antibodies have several distinguishing features. First, they are "promiscuous", that is, they interact with a number of other Mabs in an idiotypic fashion. Presumably, every antibody has the potential to interact with numerous other antibodies and so is promiscuous. However, these multiple interactions would mainly be with anti-Id directed against individual Id on the original antibody. These anti-Id would not be linked in an organized way to each other except via the Id-bearing antibody. This behavior might be referred to as "random promiscuity". In addition to this feature, super organizers appear to recognize related and possibly regulatory Id. As shown here, they bind to antibodies in a given network in an organized and predictable manner. Therefore, in contrast to the average anti-Id, super organizers display a "concerted promiscuity".

The second characteristic feature of super organizers is their heavy chain isotype. These antibodies are almost exclusvely IgM, although the significance of this heavy chain isotype restriction is not clear. Finally, super organizer antibodies often exhibit a directionality to their binding, that is, the antibody-antigen interaction does not occur equally well in both directions. In a typical case, the super organizer will bind to the Id coated onto a polyvinyl plate, but will not bind to the soluble Id. This recognition is not simply due to denaturation of the antibody by binding to the plastic plate, because the super organizers also bind to the Id on the cell surface of the relevant hybridoma in the absence of fixation.[32] In fact, this directional recognition may be a desirable feature, because it would prevent the formation of soluble antibody-antibody complexes and would still permit the anti-Id to deliver their regulatory signals at the cell surface. There are other examples of directional binding of antibodies from several laboratories, including the work of Holmberg et al.[43] In addition, Kohler's autobody does not bind to itself unless it is first attached to the ELISA plate. Finally, Geysen has described a Mab against sperm whale myoglobin which only binds to the antigen on polystyrene plates and not to soluble antigen.[33] This Mab recognizes an epitope composed of a sequence of four amino acids.

This last work raises the question, what do super organizers recognize? Most likely, these antibodies recognize amino acid sequence-related Id or epitopes. We suggest this type of recognition for two reasons. First, these antibodies are similar to the one described by Geysen where the structural requirements of the epitope are known. Second, it seems unlikely that the diverse molecules recognized by a super organizer share overall three-dimensional structure. This is especially true for JMC2 which binds to BGT, the AChR, and to several antibodies. Preliminary analysis has detected three separate regions of amino acid homology between BGT and the AChR, suggesting possible binding sites for JMC2. In the future, the antibodies EB3 and ACR24 will be sequenced to determine whether similar stretches of amino acids are present in their V regions.

It is still possible that the overall molecular conformation contributes to the shared epitope. Victor-Kobrin et al.[34] have examined the basis for shared idiotypy among antibodies which appear to have properties very similar to our own. They determined that antibodies against diverse antigens, but with shared Id, are all encoded by the heavy chain V region genes of the J558 family. They concluded that the cross-reactive Id is created by conformational determinants, because sequence comparisons could not account for the common Id. Thus, sequence and conformational deteminants can both contribute to the formation of a shared Id.

In light of the above discussion, do super organizers really represent a special subclass of anti-Id? These antibodies do share certain features (binding properties, etc.) which distinguish them from most anti-Id. As discussed earlier, the super organizers impose a hierarchy on the Id connections within a given network. Others have speculated on the importance of a hierarchy in Id networks[35,36] and we have shown here how these interactions can be arranged and organized. However, it is not necessary to propose a separate subclass of anti-Id to account for these findings.

There is one final piece of evidence which suggests that super organizers are a special subset of anti-Id. During the ontogeny of the antibody repertoire, the first B cells to be identified produce antibodies which are very similar to those described here. Holmberg reported that B cells secreting the so-called "promiscuous" antibodies were common in neonatal tissues, whereas they were rarely found in adults.[37] Kearney and Vakil[38] have published similar findings based on even earlier time points. From these studies, it can be concluded that many of the earliest B cells produce antibodies which are broadly reactive and show a high degree of Id connectivity. In the adult, these antibodies comprise a much smaller part of the repertoire, suggesting an age-related alteration in the B-cell compartment. There could be many reasons for the differential frequencies of promiscuous antibodies in the neonatal and adult repertoire. Somatic mutation of immunoglobulin genes occurs at a high rate in B cells. Over time, some of these mutations might result in the progressive loss of promiscuity for a super organizer antibody. Alternatively, super organizers may change very little from the neonate to the adult and are simply diluted out by less promiscuous antibodies. This latter explanation would appear to require separate B-cell lineages. In fact, separate lineages of B cells, based on the Ly-1 marker, have been described.[39] Interestingly, autoantibodies and IgM class immunoglobulins are prominent among the Ly-1 B-cell subpopulation.[40] Further studies will determine whether super organizers belong to the Ly-1 B-cell population.

Given the unique characteristics of the super organizer antibodies and their developmental priority, it is possible that these antibodies comprise a separate subclass of anti-Id. Perhaps this question will be answered by functional studies designed to examine the regulatory role of these antibodies.

B. Regulation by Super Organizers

Because super organizer antibodies link the immune responses to diverse antigens, it is important to determine whether they can regulate each of the individual immune responses. Some evidence already exists indicating that they can participate in each system. Victor-Kobrin et al.[34] have demonstrated that anti-Id recognizing a cross-reactive Id found on antibodies of different specificities can suppress the immune response to both antigens (DEX and NP). Vakil and Kearney[41] have also shown that administration of a single anti-Id which participates in two different antigen systems can modulate both immune responses. The timing of the anti-Id treatment was crucial for these experiments. Neonatal mice were used in the first set of studies, whereas the latter report demonstrates idiotypic regulation in both neonatal and adult mice.

Under natural conditions of infection and immunity, it is not clear to what extent idiotypic regulation occurs. Naturally occurring anti-Id have been observed and anti-anti-Id can be produced as well. But how far does this cascade of Id (Ab_1), anti-Id (Ab_2), etc. extend? If idiotypic regulation is degenerate, the cascade will end when the concentration of, say, Ab_n decreases below the level which will stimulate the production of Ab_{n+1}. While degeneracy may occur, there is little room for fine tuning or ultimate control of the production of a given Id. On the other hand, the Id cascade may cycle back on itself because of the internal image concept to provide network regulation. By this concept, the anti-anti-Id or Ab_3 would resemble the original Id or Ab_1. This arrangement of Id could lead to the formation of a closed circuit and, thereby, limit the vertical degeneracy of the response. However, upon immunization, multiple Id cycles or, more accurately, mini-networks[36] would be initiated leaving the problem of horizontal degeneracy.

We suggest that super organizers serve to limit both the vertical and horizontal degeneracy of the Id cascade. In our model, super organizers would be the glue that binds the mini-networks together into a cohesive immune response. Functionally, super organizers might be thought of as the "shock absorbers" of the immune system because they dampen the reverberations in Id networks set in motion by the introduction of antigen. This functional behavior of super organizers may be important for averting the development of an autoimmune response, because, as we have shown here, the immune responses to self- and nonself-antigens are inextricably linked. A failure to adequately control the horizontal spread of Id interactions could lead to the initiation of autoimmunity via the model outlined in this chapter.

For the future, it will be important to better understand the normal operation of Id networks and their relationship to disease states. This knowledge may not only allow us to specifically suppress autoimmunity, but also to enhance the immune response to clinically important pathogens.

ACKNOWLEDGMENTS

I would like to thank Dr. Ronald Bradley for his comments on the chapter, Dr. Meenal Vakil for helpful discussions and providing some of the antibodies used here, and Ms. Jackie Boswell for preparing the manuscript. This work was supported by grants from the Muscular Dystrophy Association, the National Multiple Sclerosis Society (RG 1734-A-1), and the Alabama Chapter of the Myasthenia Gravis Foundation.

REFERENCES

1. **Jerne, N. K.,** Towards a network theory of the immune system, *Ann. Inst. Pasteur Immunol.,* 125C, 373, 1974.
2. **Eichmann, K. and Rajewsky, K.,** Induction of T and B cell immunity by anti-idiotypic antibody, *Eur. J. Immunol.,* 5, 661, 1975.
3. **Augustin, A. and Cosenza, H.,** Expression of new idiotypes following neonatal idiotypic suppression of a dominant clone, *Eur. J. Immunol.,* 6, 497, 1976.
4. **Cazenave, P.-A.,** Idiotypic-anti-idiotypic regulation of antibody synthesis in rabbits, *Proc. Natl. Acad. Sci. U.S.A.,* 74, 5122, 1977.
5. **Rajewsky, K. and Takemori, T.,** Genetics, expression, and function of idiotypes, *Annu. Rev. Immunol.,* 1, 569, 1983.
6. **Metzger, D., Miller, A., and Sercarz, E.,** Sharing of an idiotypic marker by monoclonal antibodies specific for distinct regions of hen lysozyme, *Nature,* 287, 540, 1980.

7. **Hiernaux, J. and Bona, C.**, Shared idiotypes among monoclonal antibodies specific for different immunodominant sugars of lipopolysaccharide of different gram-negative bacteria, *Proc. Natl. Acad. Sci. U.S.A.*, 79, 1616, 1982.

8. **Hornbeck, P. V. and Lewis, G. K.**, Idiotype connectance in the immune system, *J. Exp. Med.*, 161, 53, 1985.

9. **Drachman, D.**, Myasthenia gravis, *N. Engl. J. Med.*, 293, 136, 1978.

10. **Lindstrom, J. and Dau, P.**, Biology of myasthenia gravis, *Annu. Rev. Pharmacol. Toxicol.*, 20, 337, 1980.

11. **Lindstrom, J.**, An assay for antibodies to human acetylcholine receptor in serum from patients with myasthenia gravis, *Clin. Immunol. Immunopathol.*, 7, 36, 1977.

12. **Monnier, V. M. and Fulpius, B. W.**, A radioimmunoassay for the quantitative evaluation of anti-human acetylcholine receptor antibodies in myasthenia gravis, *Clin. Exp. Immunol.*, 29, 16, 1977.

13. **Dwyer, D. S., Bradley, R. J., Urquhart, C. K., and Kearney, J. F.**, Naturally occurring anti-idiotypic antibodies in myasthenia gravis patients, *Nature*, 301, 611, 1983.

14. **Dwyer, D. S., Bradley, R. J., Oh, S. J., and Kearney, J. F.**, Idiotypes in myasthenia gravis, in *Idiotypy in Biology and Medicine*, Kohler, H., Urbain, J., and Cazenave, P.-A., Eds., Academic Press, New York, 1984, 347.

15. **Dwyer, D. S., Vakil, M., and Kearney, J. F.**, Idiotypic connectivity and a possible cause of myasthenia gravis, *J. Exp. Med.*, 164, 1310, 1986.

16. **Dwyer, D. S., Vakil, M., Bradley, R. J., Oh, S. J., and Kearney, J. F.**, A possible cause of myasthenia gravis: idiotypic networks involving bacterial antigens, *Ann. N.Y. Acad. Sci.*, in press.

17. **Kearney, J. F., McCarthy, M., Stohrer, R., Benjamin, W. H., and Briles, D. E.**, Induction of germline anti-α 1,3 dextran antibody responses in mice by members of the *Enterobacteriaceae* family, *J. Immunol.*, 135, 3468, 1985.

18. **Plotz, P. H.**, Autoantibodies are anti-idiotype antibodies to antiviral antibodies, *Lancet*, ii, 824, 1983.

19. **Saida, T., Saida, K., Lisak, R. P., Brown, M. J., Silberberg, D. H., and Asbury, A. K.**, *In vivo* demyelinating activity of sera from patients with Guillain-Barre syndrome, *Ann. Neurol.*, 11, 69, 1982.

20. **Abramsky, O., Teitelbaum, D., Webb, C., and Arnon, R.**, Cell-mediated immunity to neural antigens in idiopathic polyneuritis and myelo-radiculitis: clinical-immunological classification of the nervous system autoimmune demyelinating disorders, *Neurology*, 25, 1154, 1975.

21. **Guilbert, B., Dighiero, G., and Avrameas, S.**, Naturally occurring antibodies against nine common antigens in normal humans, *J. Immunol.*, 128, 2779, 1982.

22. **Dighiero, G., Lymberi, P., Holmberg, D., Lundquist, I., Coutinho, A., and Avrameas, S.**, High frequency of natural autoantibodies in normal newborn mice, *J. Immunol.*, 134, 765, 1985.

23. **Lymberi, P., Dighiero, G., Ternynck, T., and Avrameas, S.**, A high incidence of cross-reactive idiotypes among murine natural autoantibodies, *Eur. J. Immunol.*, 15, 702, 1985.

24. **Bona, C., Finley, S., Waters, S., and Kunkel, H. G.**, Anti-immunoglobulin antibodies, *J. Exp. Med.*, 156, 986, 1982.

25. **Chen, P. P., Fong, S., Houghten, R. A., and Carson, D. A.**, Characterization of an epibody, *J. Exp. Med.*, 161, 323, 1985.

26. **Kang, C.-Y. and Kohler, H.**, Immunoglobulin with complementary paratope and idiotope, *J. Exp. Med.*, 163, 787, 1986.

27. **Dwyer, D. S. and Schoenbeck, S.**, Anti-idiotypic regulation of the immune response against the acetylcholine receptor, in *UCLA Symp. Proc., Vol. 40: Immune Regulation by Characterized Polypeptides*, Goldstein, G., Bach, J. F., and Wigzell, H., Eds., Alan R. Liss, New York, 1987, 607.

28. **Raftery, M. A., Hunkapiller, M. W., Strader, C. D., and Hood, L. E.**, Acetylcholine receptor: complex of homologous subunits, *Science*, 208, 1454, 1980.

29. **Noda, M., Takahashi, H., Tanabe, T., Toyosato, M., Kikyotani, S., Hirose, T., Asai, M., Takashima, H., Inayama, S., Mujata, T., and Numa, S.**, Primary structures of β and δ subunit precursors of *Torpedo californica* acetylcholine receptor deduced from cDNA sequences, *Nature*, 301, 251, 1983.

30. **Berman, P. W. and Patrick, J.**, Linkage between the frequency of muscular weakness and loci that regulate immune responsiveness in murine experimental myasthenia gravis, *J. Exp. Med.*, 152, 507, 1980.

31. **Nossal, G. J. V. and Pike, B. L.**, Clonal anergy: persistence in tolerant mice of antigen-binding B lymphocytes incapable of responding to antigen to mitogen, *Proc. Natl. Acad. Sci. U.S.A.*, 77, 1602, 1980.

32. **Dwyer, D. S., Vakil, M., and Kearney, J. F.**, Directional binding of anti-idiotypic antibodies, in preparation.

33. **Geysen, H. M.**, Antigen-antibody interactions at the molecular level: adventures in peptide synthesis, *Immunol. Today*, 6, 152, 1985.

34. **Victor-Kobrin, C., Manser, T., Moran, T. M., Imanishi-Kari, T., Gefter, M., and Bona, C. A.**, Shared idiotopes among antibodies encoded by heavy-chain variable (V_H) gene members of the J558 V_H family as basis for cross-reactive regulation of clones with different antigen specificity, *Proc. Natl. Acad. Sci. U.S.A.*, 82, 7696, 1985.

35. **Kohler, H., Levitt, D., and Bach, M.,** A non-galilean view of the immune network, *Immunol. Today,* 2, 58, 1981.
36. **Paul, W. E. and Bona, C.,** Regulatory idiotopes and immune networks: a hypothesis, *Immunol. Today,* 3, 230, 1982.
37. **Holmberg, D., Forsgren, S., Ivars, F., and Coutinho, A.,** Reactions among IgM antibodies derived from normal, neonatal mice, *Eur. J. Immunol.,* 14, 435, 1984.
38. **Kearney, J. F. and Vakil, M.,** Functional idiotype networks during B-cell ontogeny, *Ann. Inst. Pasteur Immunol.,* 137C, 77, 1986.
39. **Hayakawa, K., Hardy, R. R., Herzenberg, L. A., and Herzenberg, L. A.,** Progenitors for Ly-1 B cells are distinct from progenitors for other B cells, *J. Exp. Med.,* 161, 1554, 1985.
40. **Hayakawa, K., Hardy, R. R., Honda, M., Herzenberg, L. A., Steinberg, A. D., and Herzenberg, L. A.,** Ly-1 B cells: functionally distinct lymphocytes that secrete IgM autoantibodies, *Proc. Natl. Acad. Sci. U.S.A.,* 81, 2494, 1984.
41. **Vakil, M. and Kearney, J. F.,** Functional characterization of monoclonal auto-anti-idiotype antibodies isolated from the early B cell repertoire of BALB/c, *Eur. J. Immunol.,* 16, 1151, 1986.
42. **Dwyer, D. S., Kearney, J. F., Bradley, R. J., Kemp, G. E., and Oh, S. J.,** Interaction of human antibody and murine monoclonal antibody with muscle acetylcholine receptor, *Ann. N.Y. Acad. Sci.,* 377, 143, 1981.
43. **Holmberg, D.,** Personal communication.

Chapter 5

ANTI-IDIOTYPE ANTIBODIES IN MYASTHENIA GRAVIS

A. K. Lefvert

TABLE OF CONTENTS

I. INTRODUCTION

The network theory as hypothesized by Jerne[1] assumes the ability of anti-idiotype (anti-Id) antibodies both to suppress and to enhance specific immune responses. These properties of anti-Id have been amply demonstrated in various experimental systems,[2] but, so far, there is little evidence for a functional role of autologous anti-Id for regulation of idiotype (Id) expression and modulation of the immune response in human diseases. Abnormal immune regulation has been suggested to contribute to the development and perpetuation of autoimmune disorders, and in human autoimmune diseases, autologous anti-Id have been demonstrated in myasthenia gravis,[3-11] in systemic lupus erythematosus,[12] in insulin-dependent diabetes,[13,14] and in autoimmune disorders of the thyroid gland.[15]

This paper will review our studies of anti-Id in human myasthenia gravis and focus on evidence for interactions between different antibody species and on the indications we have that such interactions may play a role for the immune regulation of the disease.

II. AUTOANTIBODIES RELATED TO THE ACETYLCHOLINE RECEPTOR

The receptor antibodies in the human disease are polyclonal.[18] The antibodies differ with regard to binding sites on the receptor, isoelectric point, and reactivity with anti-Id antibodies. There is now abundant evidence that all antibodies which bind to the acetylcholine receptor can disturb the neuromuscular function.[17] Elimination of antibodies by plasmapheresis or lymph drainage leads to improvement, and retransfusion of antibody-containing IgG to worsening of symptoms. This can be documented both as clinical deterioration and as specific electrophysiological changes.[18] The disease can be also be passively transferred to animals both by IgG from myasthenic patients and by monoclonal murine antireceptor antibodies.[17]

However, despite the clear evidence that antibodies binding to the receptor can mediate neuromuscular symptoms, there are many questions regarding their role in human disease that remain to be solved. There is only a small positive correlation between the amount of such antibodies in serum and the clinical severity of symptoms in the individual patient.[17] Antibodies can be found in high concentrations in patients in complete remission, patients who have no electrophysiological or clinical signs of disturbed neuromuscular function.[19] Antibodies are also found in all newborn infants of myasthenic mothers, usually in higher concentrations than in the mother, but only a minority of these newborns have transient symptoms of muscular weakness.[20] Lastly, antibodies binding to the receptor are found not only in clinically overt myasthenia gravis, but in healthy first-degree relatives of myasthenia patients,[21,22] and also in other diseases, patients with primary biliary cirrhosis,[23,24] patients with certain other autoantibodies,[24] thymoma patients,[25] in certain hematological disorders,[26] and in patients after bone marrow grafting.[6,27,28] Thus, antibodies binding to the acetylcholine receptor, hitherto regarded as pathogenetic and specific for myasthenia gravis, are found in disease states not accompanied by muscular weakness. Clearly, some factors other than the mere presence of antireceptor antibodies must be important for the development of disturbances of neuromuscular function. Such mechanisms may be individual variations in the multiple adaptive and protective factors operating to maintain neuromuscular transmission by regulation of the rate of receptor degradation and synthesis following antibody binding in vivo. Other factors involved may be differences in the antibody species present and the effect of immune regulation on antibody expression and pathogenicity.

In addition to antibodies binding to the receptor, there is now evidence for the spontaneous occurrence of several species of antibodies bearing receptor antibody-related idiotopes and of anti-Id antibodies directed against idiotypic determinants on the receptor antibodies. Such antibodies may obviously play a role for the immune regulation of the disease. Experimental evidences supporting this theory will be discussed in the following sections.

A. Determination of Receptor-Related Autoantibodies

Receptor antibody activity is determined as antibody binding to a partially purified human skeletal muscle preparation, trace labeled with ^{125}I-α-bungarotoxin.[16] Antibodies directed exclusively against the toxin binding region of the receptor are measured using a modification of this assay.[16]

Receptor antibody-related Id in sera and cell culture supernatants are measured by binding to six different BALB/c monoclonal anti-Id antibodies in an ELISA. These monoclonal anti-Id antibodies are designed as A121, 3, 8, 24, 41, and 45, respectively.[8-10] A121 was raised against the purified heavy chain from the IgG3-Fab fragment from a patient with receptor antibody activity exclusively against the transmitter binding region of the receptor.[29] This particular anti-Id antibody bears an internal image of the transmitter binding region of the receptor, since it binds cholinergic ligands. It is also able to initiate the synthesis of receptor antibodies in experimental animals.[30] This anti-Id recognizes a recurrent idiotope found on immunoglobulins in about 60% of patients with myasthenia.[29] The other anti-Id antibodies were all raised against affinity-purified receptor antibodies from patients with myasthenia.[16] A13, 8, and 41 bind to immunoglobulins in 27 to 38% of patients in an ELISA, while inhibition of receptor antibody binding to antibody by anti-Id can be demonstrated in 5 to 13% of patients. A124 and 45 are probably directed against framework determinants, since they cannot be demonstrated to inhibit binding of receptor antibody to receptor. Direct binding in ELISA can be demonstrated in 14% of myasthenia patients.[8,10,31]

Anti-Id antibodies are measured by binding to BALB/c murine monoclonal antireceptor antibodies. These monoclonals are raised against the purified receptor of *Torpedo marmorata* and selected primarily because of their strong cross reactivity with human receptor.[6,7] They are designed as TR63, 94, and 105, respectively. TR63 is directed against the toxin-binding region of the human receptor and inhibits up to 56% of toxin binding, using 100 μg of TR63, 0.5 pmol of receptor (measured as toxin-binding sites), and a five-fold excess of α-bungarotoxin. TR94 and 105 are directed against peripheral parts of the receptor and do not inhibit toxin binding. TR94 binds 8.8×10^{-14} and TR105 15.8×10^{-14} mol toxin-receptor complex per 100 μg purified antibody.[6,7] Inhibition experiments performed by preincubation of purified F(ab')$_2$ fragments from one monoclonal receptor antibody, with receptor before addition of another monoclonal, showed that none of these antireceptor antibodies inhibited the binding of one another to receptor.[7]

B. Specificity of Binding of Human Autoantibodies to the Monoclonal Anti-Id and Antireceptor Antibodies

Mouse monoclonal antibodies were used to define the human immunoglobulins. The specificity of the binding of the human antibodies to the monoclonal antibodies was ascertained as described.[6,7] Briefly, the structures mediating binding are localized to the F(ab')$_2$ parts of the antibodies, the specificities make interactions with mouse or human alloepitopes unlikely, and there is no interference with potentially cross-reacting agents such as sDNA, phospholipids, or a bacteria lipopolysaccharide. It may, therefore, be concluded that the murine monoclonal antibodies recognize unique epitopes on myasthenia immunoglobulins, probably idiotopes which resemble structures on the acetylcholine receptor and on anti-Id antibodies, respectively.

III. PROPERTIES OF ANTI-ID ANTIBODIES

Four anti-Id antibodies from four patients have been isolated by affinity chromatography on immobilized monoclonal antireceptor antibodies. The isotype of three of them was IgG-kappa and of one IgM-kappa. One IgG-kappa was purified by its binding to TR63 and obtained in amounts sufficiently high to allow some characterization. This antibody showed

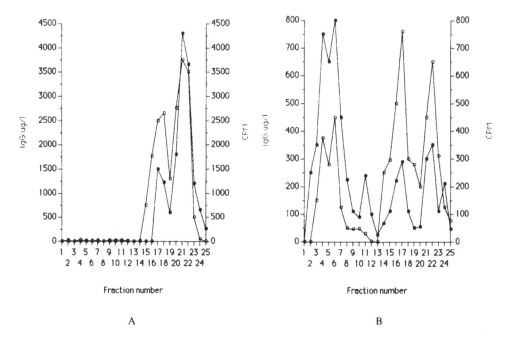

FIGURE 1.   (A) Sucrose density gradient centrifugation of a radiolabeled monoclonal human IgG receptor antibody together with a human IgG myeloma protein. (○): IgG μg/ℓ; (●): CPM. (B) Sucrose density gradient centrifugation of a radiolabeled monoclonal human IgG receptor antibody together with a human monoclonal IgG anti-Id antibody binding to TR63 from the same patient. (○): IgG μg/ℓ; (●): CPM.

a restricted heterogeneity. Three bands within the pH range of 7.83 to 8.05 were demonstrated with isoelectric focusing followed by immunofixation and visualization by biotin-avidin-peroxidase. This particular anti-Id antibody did not inhibit the binding of serum-containing receptor antibodies from the same patient to receptor, although the concentration of anti-Id may well have been too low. Alternatively, the affinity of receptor antibodies for receptor may be higher than that of the anti-Id for the receptor antibody, or the anti-Id may bind to idiotopes outside the antigen binding region.[6] Preliminary results using monoclonal antibodies from lymphoblastoid lines obtained by Epstein-Barr virus (EBV-)transformed peripheral blood B lymphocytes from patients with myasthenia, give supportive evidence for the anti-Id nature of these immunoglobulins binding to the murine monoclonal antireceptor antibodies.[8] Figure 1A shows the sucrose density gradient pattern of a purified, radiolabeled human monoclonal IgG1-kappa with antibody activity against the acetylcholine receptor when preincubated with a control human monoclonal antibody (IgG1-lambda myeloma protein) and subjected to centrifugation on a continuous sucrose density gradient (5 to 20% with a bottom layer of 40% sucrose). No complex formation occurs; the radiolabeled antibody is found in the fractions expected for 7S immunoglobulins. Figure 1B shows the same monoclonal antireceptor antibody when preincubated with a monoclonal anti-Id antibody (IgG1-lambda) from the same patient, selected by its binding to TR63. A variety of complexes with different sedimentation constants is formed. The complex formation could not be inhibited by pooled normal IgG. This must be considered evidence both for the anti-Id nature of the immunoglobulins selected by the reaction with monoclonal murine antireceptor antibodies, and also for possibilities of interactions in vivo between such antibodies and receptor antibodies.

IV. PREVALENCE OF ANTI-ID ANTIBODIES

The prevalence of anti-Id antibodies reacting with one or more of the different monoclonal

Table 1
PREVALENCE OF ANTI-ID ANTIBODIES REACTING WITH ONE OR MORE OF THE MONOCLONAL ANTIRECEPTOR ANTIBODIES TR63, 94, AND 105 AND OF ANTIBODIES AGAINST THE ACETYLCHOLINE RECEPTOR IN DIFFERENT CLINICAL STAGES OF MYASTHENIA GRAVIS

	Anti-Id (%)	Anti-receptor (%)
Complete remission	50	75
Ocular myasthenia	30	60
Mild generalized	69	96
Severe generalized	61	100
Early disease (<1 year)	96	93
Late disease (>5 years)	62	92

Note: The mean + 2SD of a normal population was used as the cutoff limit.

antireceptor antibodies and of antibodies binding to the acetylcholine receptor is shown in Table 1. The clinical classification was done according to Osserman-Oosterhuis.[32] Patients with myasthenia in Stage IIA (mild, generalized) were considered as having mild disease, and patients in Stages IIB (moderately severe, generalized), III (early severe), and IV (late severe) as having severe disease.

Patients who have been in complete remission for a long time do not, as perhaps may have been expected, tend to have a higher incidence of anti-Id than others. The most remarkable and unexpected result was the high prevalance of anti-Id in early disease. These patients also have significantly higher anti-Id concentrations than any other groups.[6]

Generally, the concentrations of receptor antibody activity and of the various Id correlate well, whereas there is — at least in some patient groups — an inverse relation between these antibody species and anti-Id. Such cases may be considered evidence for interactions between the different antibody species and will be discussed in the following sections.

V. ANTI-ID ANTIBODIES DURING DEVELOPMENT OF THE DISEASE

The high incidence and high concentration of anti-Id in early disease prompted us to do a special investigation of the autoantibody pattern during this period. Thirteen patients with myasthenia of acute onset and with a duration of subjective symptoms of less than 1 month were selected for this investigation. In addition, we had the opportunity of following the appearance of specific antibodies in two patients who developed myasthenia after bone marrow grafting. Table 2 shows the prevalence of receptor antibody activity of IgM and IgG class and of anti-Id antibodies of both IgG and IgM class in the 13 patients at the first examination (3 days to 4 weeks after start of symptoms) as compared to that found in long-standing myasthenia. The incidence of IgM receptor antibodies is very high, as is that of anti-Id. About half of the patients had anti-Id either exclusively of IgM class or of both IgG and IgM class; the rest had IgG anti-Id.

Table 2
PREVALENCE OF ANTI-ID REACTING
WITH ONE OR MORE OF THE
MONOCLONAL ANTIRECEPTOR
ANTIBODIES TR63, 94, AND 105 AND OF
ANTIBODIES AGAINST THE
ACETYLCHOLINE RECEPTOR IN
MYASTHENIA OF RECENT ONSET AND
IN LONG-STANDING DISEASE

Duration of symptoms	IgG receptor antibodies	IgM receptor antibodies	Anti-Id antibodies
<4 Weeks	11/13	11/13	13/13
>5 Years	73/79	10/79	49/79

In a few patients, it was possible to do serial determinations of autoantibodies during the development of the disease in untreated patients. In these patients, a very consistent pattern was found. Initially, high levels of anti-Id were present, in most patients together with antireceptor antibodies of IgM class. Both these antibody species usually decreased markedly as IgG receptor antibody activity appeared. In most patients, IgG receptor antibodies continued to rise and IgM receptor antibodies disappeared, while the concentration of anti-Id diminished and eventually stabilized on a lower level. Figure 2 shows the autoantibody pattern in such a patient. She was a 20-year-old female who developed acute symptoms of neuromuscular weakness about 1 week after a mild upper respiratory infection. At the time of the first examination, she had no IgG receptor antibodies, but high levels of IgM receptor antibodies and anti-Id binding to TR63. As the disease progressed, IgG receptor antibodies appeared and the other autoantibody species decreased in concentration.

Two patients who had received bone marrow grafts could be followed even before the appearance of clinical symptoms of myasthenia. Figure 3 shows the autoantibody pattern in a teenage girl who received a graft from her HLA-A, B, C, and D identical brother for treatment of severe aplastic anemia of unknown origin.[33] She developed a chronic graft-vs.-host disease (GVHD) and 27 months after the transplantation, severe myasthenia gravis. At that time, all her peripheral and bone marrow cells carried Y chromosomes, indicating a complete takeover of the donor cells. Anti-Id antibodies were present already before the grafting in this patient. A suggested peak of anti-Id preceded the receptor antibodies of IgG class that were demonstrated a few months after grafting. During the following months, when the patient had no clinical symptoms of myasthenia, there was a suggested inverse relationship between the anti-Id antibodies and the IgG receptor antibodies. At the time of start of myasthenic symptoms, IgG receptor antibodies showed a marked rise, followed by a second peak of anti-Id antibodies. The concentration of IgG receptor antibodies remained high until her death, but anti-Id and IgM receptor antibodies decreased markedly. In this case, there is a question of familial disposition for myasthenia, since both the donor, her clinically healthy brother, and her father had abnormal single fiber electromyographic recordings with increased jitter. None of them had antibodies reacting with the acetylcholine receptor. The donor also had an anti-Id reacting with the same monoclonal antireceptor antibody (TR63) as that found in the patient. The IgG fraction from the donor serum was able to inhibit 60% of IgG receptor antibody binding to receptor in the patient, indicating anti-Id antibodies in the donor that were directed against the antigen binding region of the receptor antibodies.

The second patient who developed myasthenia after bone marrow grafting was a young

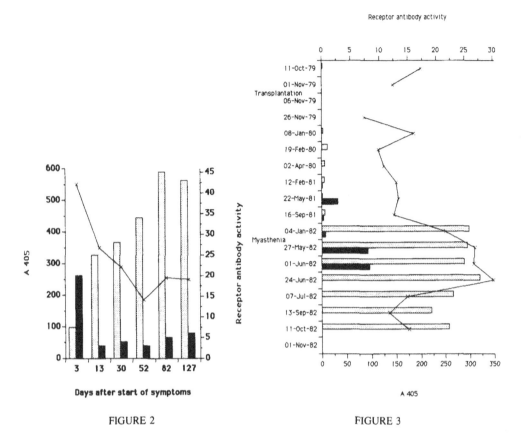

FIGURE 2

FIGURE 3

FIGURE 2.   Serum IgG receptor antibody activity, lightly shaded (percent inhibition of toxin binding); IgM receptor antibody activity, darker shaded (moles $\times$ $10^{-10}/\ell$ serum), and anti-Id antibody binding to TR63 (x) (normal population: mean + 2SD = 123 milliabsorbance units) in a patient with early myasthenia.

FIGURE 3.   Serum IgG receptor antibody activity, lightly shaded (moles $\times$ $10^{-9}/\ell$); IgM receptor antibody activity, darker shaded (moles $\times$ $10^{-9}/\ell$), and anti-Idiotype antibodies binding to TR63 (x) (normal population: mean + 2SD = 119 milliabsorbance units) in a patient with aplastic anemia who developed myasthenia gravis after bone marrow transplantation.

infant with severe combined immune deficiency (Figure 4).[34] She received the graft at 6 months of age from her mother and developed a chronic GVHD and signs of severe myasthenia 7 months following the transplantation. Gm analysis of her immunoglobulins indicated a complete takeover of the donor cells. Both her parents were clinically healthy and none of them had any autoantibodies related to the acetylcholine receptor. In this infant who had no autoantibodies related to the acetylcholine receptor before the transplantation, there was also a clear priority of the anti-Id response, which appeared already a few months after grafting. Very low concentrations of IgG receptor antibodies were detectable at the same time. When she developed clinical signs of severe myasthenia, the anti-Id disappeared and IgG receptor antibodies rose to high concentrations. In these two cases, IgM receptor antibodies did not precede the IgG antibodies. This is in contrast to the pattern seen in early spontaneous myasthenia.

A priority of the anti-Id response has been described, also, in other systems after immunization with antigen.[35] Since it is such a consistent finding in early myasthenia, it suggests that the expression of anti-Id is critical in the induction phase of the disease.

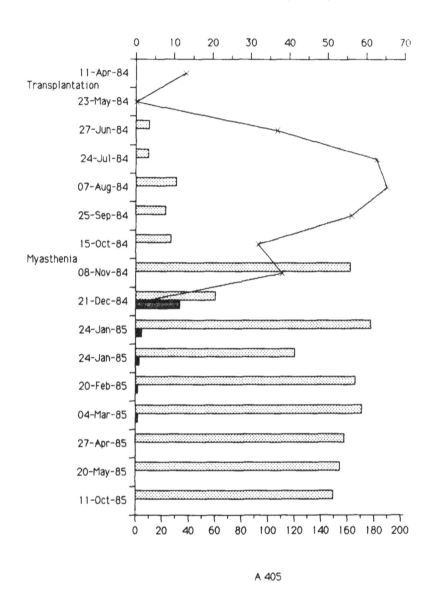

A 405

FIGURE 4.   Serum IgG receptor antibody activity, lightly shaded (moles × $10^{-9}/\ell$); IgM receptor antibody activity, darker shaded (moles × $10^{-9}/\ell$), and anti-Idiotype antibodies binding to TR63 (x) (normal population: mean + 2SD = 109 milliabsorbance units) in an infant with severe combined immunodeficiency who developed myasthenia gravis after bone marrow transplantation.

VI. ANTI-ID ANTIBODIES IN TRANSIENT MYASTHENIA GRAVIS SYNDROMES

There are two instances in which one can talk about a transient myasthenia gravis syndrome: one is neonatal myasthenia[36] and the other is the penicillamine-induced disease.[37] Typically, in such cases the myasthenic symptoms disappear completely a few months after birth or cessation of treatment, respectively. In neither case is there an increased tendency to develop the disease later in life.

A. Neonatal Myasthenia

Neonatal myasthenia is a disorder that affects only 12% of newborns of myasthenic mothers.[36] There is a paradoxical dissociation between antibody levels in the newborn and disease, indicating that other factors than the mere presence of antireceptor antibodies transferred from the mother must be of importance. We have earlier presented evidence for receptor antibody synthesis in the affected infants.[20] These conclusions were based on two experimental findings: (1) the half-life of the receptor antibodies was longer than expected in two out of three affected children (27 to 91 days) and blood exchange did not alleviate the symptoms in two of the infants, and (2) affected children had receptor antibodies bearing other idiotopes than those found in the mother. Healthy children of myasthenic mothers did not show this difference in Id repertoire and the half-lives of their receptor antibodies were frequently shorter than that of normal IgG (4 to 11 days).[20]

The prevalence and concentration of receptor antibodies and of anti-Id antibodies were determined in 14 children. Three of them had neonatal myasthenia. All children had the same or slightly higher antibody concentrations than their mothers at birth. The three affected children had comparatively high concentrations of IgG receptor antibodies — 34, 22, and 28 nmol/ℓ, respectively. Two healthy children had antibody concentrations in the same range, but the mean concentration in this group was lower than in affected children (mean 9.5; range 0 to 28.6 nmol/ℓ). Anti-Id, however, were found in detectable amounts in 7 out of 11 healthy children, but in none of the 3 affected infants. In almost all infants, the concentration of anti-Id was lower than that found in the mother.[6] The mothers of the three affected infants had anti-Id in somewhat lower concentrations than in mothers of healthy children.[6] During the first months of life, anti-Id production was demonstrated both in healthy and affected infants. This is illustrated in Figures 5A and 5B, showing the anti-Id response in a healthy child and in a child with severe neonatal myasthenia. Neither of them had any anti-Id demonstrable at birth. The increase in anti-Id in the healthy child is accompanied by a rapid decrease in receptor antibody concentration. In the affected child, the same pattern is suggested, although the changes in concentration of the antibody species are much slower.

These findings may indicate that anti-Id play a role for the prevention and/or disappearance of neonatal myasthenia. Infants who have a relative excess of anti-Id, either transferred from their mothers or produced by themselves, seem to be less prone to develop neonatal myasthenia. The very short half-life of the receptor antibodies found in these healthy infants may be due to complex formation with anti-Id and subsequent rapid elimination. In most children, both affected and healthy, a production of anti-Id after birth can be demonstrated,[6,8] a process that may be of importance for the termination of the neonatal disease.

B. Penicillamine-Induced Myasthenia

The autoantibody pattern in a middle-age female patient with severe rheumatoid arthritis and penicillamine-induced myasthenia during and 2 months after cessation of penicillamine treatment is shown in Figure 6. During treatment, this patient had high levels of receptor antibodies and Id that reacted with five out of six anti-Id antibodies. After cessation of treatment, these antibody species had decreased markedly, but two species of anti-Id antibodies showed an increase. Thus, also in penicillamine-induced myasthenia, a switch from Id to anti-Id dominance accompanies the gradual disappearance of myasthenia symptoms and of receptor antibodies.

VII. BONE MARROW GRAFTING AND RECEPTOR-RELATED AUTOANTIBODIES

A. Autoantibodies after Bone Marrow Grafting

The incidence of myasthenia gravis after bone marrow grafting is high, at least 20 times

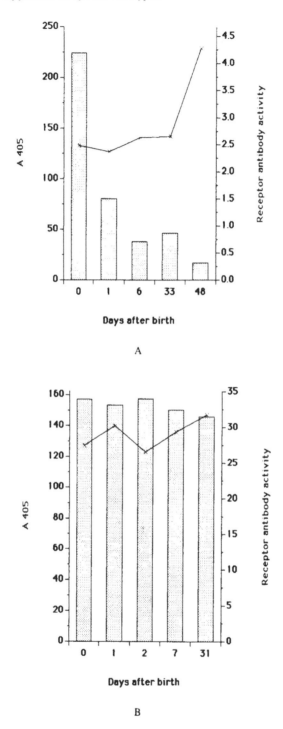

A

B

FIGURE 5.   (A) Serum IgG receptor antibody activity, shaded (moles $\times$ $10^{-9}/\ell$), and anti-Id antibodies binding to TR105 (x) (normal population: mean + 2SD = 128 milliabsorbance units) during the first 6 weeks of life in a healthy child of a myasthenic mother. (B) Serum IgG receptor antibody activity, shaded (moles $\times$ $10^{-9}/\ell$), and anti-Id antibodies binding to TR105 (x) (normal population: mean + 2SD = 125 milliabsorbance units) during the first month of life in a child with neonatal myasthenia.

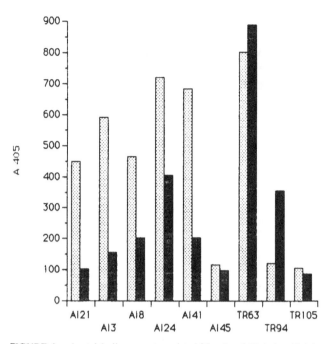

FIGURE 6. Acetylcholine receptor-related Id and anti-Id during (lightly shaded) and two months after discontinuation of treatment with penicillamine (darker shaded) in the serum of a patient with penicillamine-induced myasthenia (normal population: mean + 2SD = 120 milliabsorbance units).

that found in a normal population. Up to now, six cases of myasthenia are known to us.[33,34,38,39] This is most probably an underestimation, since the treatment these patients get should be very effective in suppressing symptoms of mild disease. Interestingly, a similar syndrome has been described in a dog who received transplantation of fetal hematopoetic cells.[40] Most patients described so far have also had a chronic graft-vs.-host reaction when they developed signs of myasthenia. The diagnoses in the patients known to us have been severe aplastic anemia (two cases of idiopathic aplastic anemia; one case of Diamond-Blackfan congenital aplastic anemia), severe combined immunodeficiency (one case) and acute myeloid leukemia (AML) (one case). Various autoantibodies have been described to be rather common after bone marrow grafting, usually in combination with chronic GVHD.[41] We have looked at the prevalence of receptor-associated autoantibodies in 51 patients after bone marrow grafting. None of the patients had clinical signs of disturbed neuromuscular function. Receptor-associated antibodies were found in 21 patients (40%). The prevalence of such antibodies as related to the preoperative diagnosis, type of graft, and complications occurring after grafting is listed in Table 3. Patients transplanted because of AML show the highest incidence of such antibodies, and this group accounts for one third of all patients with antibodies (7 out of 21 patients). Patients with acute lumphocytic leukemia (ALL) have the lowest incidence of antibodies after grafting. There is also a tendency for a negative correlation between antibodies and chronic GVHD. The differences between patients with AML and other patients were not statistically significant ($p = 0.12$), neither was there any significant difference in antibody incidence between patients with and without chronic GVHD ($p = 0.13$).

There was, however, a highly significant difference in antibody incidence depending on the type of graft. All 4 patients with syngeneic grafts (2 identical twins transplanted because of AML and chronic myelocytic leukemia [CML] and 2 autologous grafts in 2 patients with neuroblastoma and Ewing sarcoma) had antibodies, whereas only 15 out of 46 patients with allogeneic grafts developed antibodies ($p < 0.01$).

Table 3
PREVALENCE OF
AUTOANTIBODIES RELATED TO
THE ACETYLCHOLINE RECEPTOR
AFTER BONE MARROW GRAFTING

Diagnosis	Ab/examined patients
ALL	4/17 (24%)
CML	2/6 (33%)
AML	7/15 (47%)
SAA	3/9 (33%)
Complications after grafting	
Acute GVHD	3/8 (38%)
No acute GVHD	12/28 (43%)
Chronic GVHD	4/20 (20%)
No chronic GVHD	11/26 (42%)
Type of graft	
Allogeneic	15/46 (33%)
Syngeneic	4/4 (100%)

Note: ALL, acute lymphocytic leukemia; CML, chronic myelocytic leukemia; AML, acute myeloid leukemia; SAA, severe aplastic anemia; and GVHD, graft-vs.-host disease.

The autoantibodies usually appear within 1 year after the transplantation, and several antibody peaks are common. The autoantibody patterns in two patients are shown in Figures 7 and 8. The patient shown in Figure 7 was a teenage girl, transplanted because of CML, who received her graft from an identical twin. She had low concentrations of receptor antibody activity already before the transplantation, a clear priority of the anti-Id response and both receptor antibody activity and associated Id. The other patient was a teenage girl with AML who received an allogeneic graft (Figure 8). She has no demonstrable antibodies binding to the receptor, but both receptor-related Id and anti-Id. In this case, there is also evidence for a switch from one dominant Id occurring shortly after the transplantation to another 1 year later. None of these patients have any neuromuscular symptoms. The characteristics of the autoantibodies in these patients have so far been poorly studied. One major difference to the autoantibodies in spontaneous myasthenia seems to be that the anti-Id antibodies in bone marrow-grafted patients without myasthenia are comparatively often of IgM class. This implies idiotypic cross reactivity between IgG and IgM anti-Id antibodies. Another difference is the frequency of Id, as determined by reaction with our six monoclonal anti-Id antibodies. The recurrent Id characterized by its binding to A121, which is found in half of patients with myasthenia, was found in only 20% of bone marrow-grafted patients with receptor antibody activity. The patients are, however, still too few to allow definite conclusion to be made.

The occurrence of specific autoantibodies related to the acetylcholine receptor after syngeneic grafts and not — unlike other autoantibodies — positively correlated to chronic GVHD speaks strongly against a nonselective B-cell stimulation and for a more specific clonal expansion of cells producing antibodies related to the acetylcholine receptor. The nature of this specific disturbance of the immune regulation, possibly achieved by antigens or antibodies bearing epitopes capable of triggering such a response, remains to be elucidated.

B. Receptor-Related Autoantibodies in Hematological Disorders

Myasthenia gravis has until now been associated with a syndrome characterized by pure

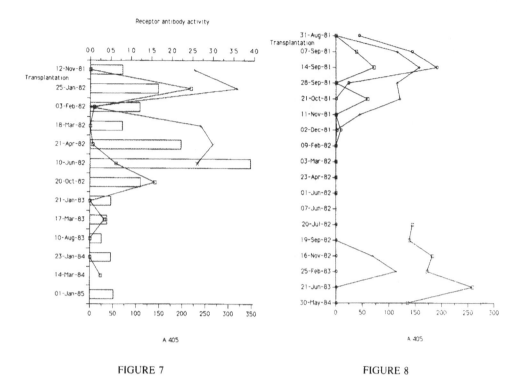

FIGURE 7   FIGURE 8

FIGURE 7.   Serum IgG receptor antibody activity, shaded (moles × 10 $^9/\ell$), an Id binding to A124 (□) and an anti-Id binding to TR105 (x) in a teenage girl with chronic myelocytic leukemia, who received a bone marrow graft from her identical twin. Normal values have been subtracted.

FIGURE 8.   Id binding to A124 (○) and A145 (□) and an anti-Id antibody binding to TR94 (x) in a teenage girl with acute myeloid leukemia who received an allogeneic bone marrow graft. Normal values have been subtracted.

red cell aplasia, thymoma, and myasthenia,[42] but not with other hematological disorders. As a corollary of the results discussed above, we proceeded to investigate the possible association of hematological disorders and acetylcholine receptor-specific autoantibodies.[26] Receptor antibody activity, associated Id, and anti-Id were determined in 56 patients with acute leukemias, in 2 patients with pure red cell aplasia, and in 3 patients with idiopathic aplastic anemia. The leukemia patients were tested before the start of treatment. The other patients were in partial or complete remissions and received no treatment. Preliminary results show antibodies in 6 of 42 patients with AML, but not in other kinds of acute leukemia. Antibodies were also found in two of three patients with aplastic anemia, but in none of the patients with pure red cell aplasia. There was a dominance of IgM class autoantibodies.

One patient with aplastic anemia in remission and no clinical evidence of muscular weakness had a relatively high receptor antibody level and was studied in more detail.[26] She had receptor antibody activity of IgG class in relatively low concentration (2.92 mol × $10^{-9}/\ell$ serum) and of IgM class in high concentration (2.70 mol × $10^{-9}/\ell$ serum). The pattern of Id and anti-Id of different isotypes is shown in Figure 9. Compared to what is found in myasthenia gravis patients, there is a marked increase in IgM class Id and anti-Id. There is also a very broad reactivity with the monoclonal anti-Id and antireceptor antibodies, something that is rarely seen in myasthenia gravis. Serum from this patient did not react with Fc γ fragments, nor with a panel of 18 murine monoclonal antibodies with irrelevent specificities.[7]

The immunoglobulins from this patient were purified by gel filtration and passively

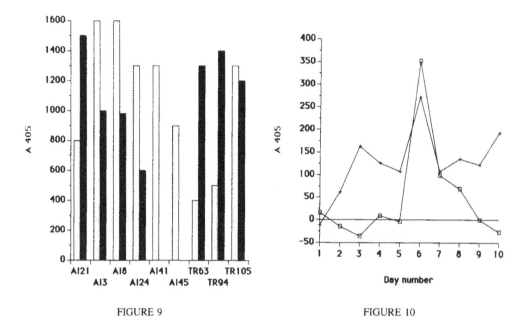

FIGURE 9   FIGURE 10

FIGURE 9.   Acetylcholine receptor-related Id and anti-idiotypic antibodies of IgG (lightly shaded) and IgM (darker shaded) class in a patient with aplastic anemia in remission. Normal values have been subtracted.

FIGURE 10.   Spontaneous release of an Id binding to A121 (+) and an anti-Id antibody binding to TR63 (□) in culture of peripheral blood mononuclear cells. Normal values have been subtracted.

transferred to three C57BL/6 mice as daily i.p. injections.[17] The mice showed no signs of neuromuscular dysfunction. After 7 days, the animals were killed, cholinergic receptor was extracted from the decapitated, skinned, and eviscerated carcasses, and the concentration of free receptor measured. The amount of receptor in these mice was compared to that of mice which had received immunoglobulins from patients with myasthenia and from normal persons, respectively. The results show that immunoglobulins from this patient with aplastic anemia and receptor antibodies primarily of IgM class are capable of binding to the receptor in vivo, and reduce the free receptor content to the same low level as that found after transfusion with immunoglobulins from patients with myasthenia gravis. This is strong evidence for a specific, high-affinity interaction with the acetylcholine receptor. The same pattern of receptor-related autoantibodies belonging to the IgM class is found in patients with primary biliary cirrhosis,[23] and immunoglobulins from these patients are also capable of reducing the receptor content in mice after passive transfer.[24]

Thus, patients with certain hematological disorders and malignancies have clones of cells, capable of producing antibodies related to the acetylcholine receptor, that are already activated. Such patients may be especially prone to develop antibodies after transplantation. This might lead to clinical disease in susceptible individuals. The theory is supported by the somewhat higher incidence of overt myasthenia gravis and of receptor-associated autoantibodies in patients transplanted because of AML and aplastic anemias.

VIII. PRODUCTION OF AUTOANTIBODIES IN CELL CULTURE SYSTEMS

Indications of interactions between different autoantibodies in myasthenia gravis are found, also, in cell culture systems.[7,8,31] Figure 10 shows the spontaneous release of an Id and an anti-Id in a culture of peripheral blood mononuclear cells from a patient with myasthenia

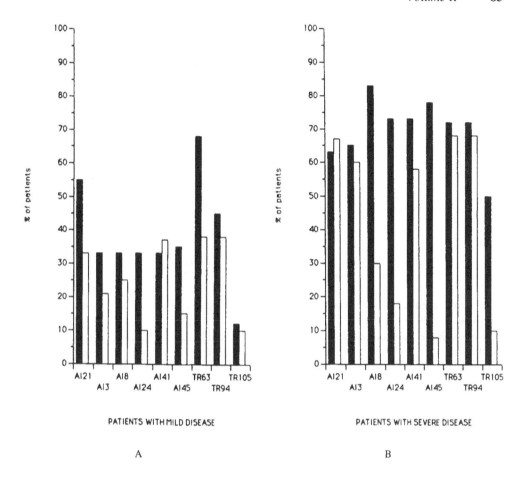

FIGURE 11.   (A) Prevalence of Id and anti-Id antibodies in sera (□) and cell culture supernatants (■) from patients with mild disease (Stages A, I, and IIA). (B) Prevalence of Id and anti-Id antibodies in sera (□) and cell culture supernatants (■) from patients with severe disease (Stages IIB, III, and IV).

gravis, set up as described.[7] Cells were cultured without change of medium and supernatants harvested at days 1 to 10. There is an early peak of Id, while the anti-Id appears later. During the following days, the Id increases in concentration and the anti-Id shows a decrease. This may be due to complex formation followed by elimination by macrophages in the system, or leading to blocking of reactive sites on the antibodies, making them nonreactive in the assays. A suggested inverse relation between Id and anti-Id in cell culture systems was seen in about 25% of cultures from 13 patients.[7,8]

Another indirect evidence for in vivo interactions between complementary antibodies are the spectra of Id and anti-Id disclosed in cell culture systems as compared to that in serum from the same patient.[31] Figures 11A and 11B show the frequencies of spontaneous production of Id and anti-Id in cell cultures and the corresponding antibody species in serum in patients with mild and severe myasthenia, respectively.[31] The prevalence of different antibody species is higher in cell culture supernatants than in serum irrespective of the clinical stage of disease. The prevalence of different species of Id as well as of anti-Id is higher in cell cultures from patients with severe rather than mild disease. The incidence of Id, but not that of anti-Id, in patients with mild disease could be somewhat increased by pokeweed mitogen stimulation, suggesting the presence of dormant clones. When the relation between the different antibody species was examined, there was a tendency for Id dominance in unstimulated cell cultures from patients with severe disease, while the opposite was true

Table 4
PREVALENCE OF ID AND ANTI-ID OF BOTH IgG
AND IgM CLASS IN PRIMARY CLONES OF
LYMPHOBLASTOID LINES OBTAINED BY EBV
TRANSFORMATION OF PERIPHERAL B
LYMPHOCYTES FROM FOUR PATIENTS WITH
MYASTHENIA GRAVIS

Patient	Id (%)	Anti-Id (%)
Patients with mild disease (3 patients; 420 clones)	3	4
Patient with severe disease (1 patient; 60 clones)	2	7

Note: Cutoff limit was mean + 4SD of values obtained with culture medium with the addition of normal IgG or IgM.

in mild myasthenia.[8] This is compatible with the assumption of a regulatory role of anti-Id antibodies for the expression of Id in the disease. The complex patterns disclosed by their reaction with the monoclonal antireceptor and anti-Id antibodies, respectively, do not allow conclusions regarding the functional role of these autoantibodies. The important pathogenetic factor should be the balance between these antibody species and not the antibody concentration. Some species of autoantibodies may be rapidly removed from serum by binding to receptor epitopes or following complex formation with complementary antibodies. It is reasonable to hypothesize that antibody species that are preferentially eliminated from serum, that is, antibodies with high affinities for antigen and/or for complementary antibodies, are important and may have a particular role in the disease process. The use of cell culture systems clearly permits the more complete detection of the different species of both Id and anti-Id than mere studies of serum samples and should, thus, be a useful tool for the investigation of autoantibody repertoire and interactions between autoantibodies.

Preliminary evidence suggesting a prominent anti-Id response in myasthenia has recently been demonstrated by us using B-lymphoblastoid lines obtained by transformation of peripheral blood B lymphocytes by Epstein-Barr virus (EBV).[43] One hundred EBV-infected B lymphocytes were set up in microtiter wells together with autologous irradiated peripheral blood mononuclear cells. Cell growth occurred in close to 100% of the wells in all experiments. These primary clones were screened for binding to the monoclonal anti-Id and antireceptor antibodies, respectively. Positive wells are subcloned and the testing repeated. In most cases, the primary clones secreted only one Ig-isotype with only one kind of light chain that was monospecific in our test systems with monoclonal antibodies, indicating a surprisingly high degree of selection already at this stage. The frequencies of Id and anti-Id of IgG and IgM class in primary clones obtained from three patients with mild myasthenia gravis and one patient with severe disease are shown in Table 4. The anti-Id-secreting clones dominate over those secreting Id, especially in the one patient with severe myasthenia.

IX. ANTI-ID ANTIBODIES IN HEALTHY FIRST-DEGREE RELATIVES OF MYASTHENIA GRAVIS PATIENTS

A genetic basis for the tendency to develop autoimmune diseases is well established on the basis of family studies and HLA associations. Relatives of patients with certain autoimmune disorders have an increased risk of developing the disease as well as an increased incidence of autoantibodies not accompanied by clinical symptoms. This is the case, also,

Table 5
RECEPTOR ANTIBODIES, A RECURRENT ID
DEFINED BY BINDING TO A121, ANTI-ID DEFINED
BY BINDING TO ANY OF THE MONOCLONAL
ANTIRECEPTOR ANTIBODIES AND PATHOLOGICAL
SINGLE-FIBER ELECTROMYOGRAPHY (SF-EMG) IN
HEALTHY RELATIVES (N = 67) TO PATIENTS WITH
MYASTHENIA GRAVIS

	Parents (%)	Siblings (%)	Children (%)
Receptor antibodies	42	59	60
Id	20	50	19
Anti-Id	27	38	43
Pathological SF-EMG	44	31	12

Note: For the antibody assays, mean + 2SD of a normal population (more than 500 persons) was used as cutoff limit. SF-EMG was considered pathological when 3 or more fiber pairs out of 20 showed increased jitter, with or without blocking.

Table 6
COMBINATIONS OF RECEPTOR ANTIBODIES, ID,
ANTI-ID, AND PATHOLOGICAL SINGLE-FIBER
EMG (SF-EMG) IN HEALTHY RELATIVES OF
PATIENTS WITH MYASTHENIA GRAVIS

Pathological findings	Parents, (n = 14 (%)	Siblings, n = 11 (%)	Children, n = 16 (%)
Receptor antibody and/or Id + anti-Id	14	55	50
Receptor antibody + pathological SF-EMG	14	18	6
Anti-Id + pathological SFEMG	7	0	0
Receptor antibody only	14	18	13
Anti-Id only	0	9	0
Pathological SF-EMG only	29	0	6
Receptor antibody and/or Id + anti-Id + pathological SFEMG	0	0	0
No abnormal finding	22	0	25

in myasthenia gravis.[21,22] Additional abnormalities in first-degree relatives include a rather high incidence of signs of disturbed neuromuscular function as determined by single-fiber electromyography (SF-EMG).[44]

The prevalence of receptor antibodies, of anti-Id, and of pathological SF-EMG in relatives is shown in Table 5, and the frequencies of the different combinations of these pathological findings in Table 6. There is a negative relation between the presence of anti-Id and pathological SF-EMG. In no case could the combination of antibody activity and/or Id, anti-Id, and pathological SF-EMG be demonstrated. This suggests that an excess of anti-Id that may form complexes with the receptor antibodies may protect the neuromuscular junction from the damaging effects of antibody binding, thereby preserving a completely normal neuro-muscular function.

Children of myasthenic mothers have a rather high incidence of receptor antibodies, Id, and anti-Id.[21,22] It is of interest to note that the particular Id and anti-Id expressed in the mother are rarely found in the children we have examined at age 4 to 20. In 75% of these children, the recurrent Id defined by A121 was found either in the mother or in the child. The same holds true for anti-Id antibodies. Only 25% of children express the same species of anti-Id as their mother. This suggests a long-lasting effect of maternal immunity on the developing immune system involving suppression of expression of certain antibody species in the child.

X. CONCLUDING DISCUSSION

I have tried to give some examples of our experimental results suggesting a functional role for Id-anti-Id interactions in myasthenia. In healthy relatives of myasthenia patients, an abnormal SF-EMG is common in those having receptor antibodies, but is never seen in combination with receptor antibodies and anti-Id. This suggests that the presence of anti-Id that bind receptor antibodies protect the acetylcholine receptor from the effects of antibody binding, thereby preserving a completely normal neuromuscular function.

A protective effect of anti-Id is also suggested in transient myasthenia gravis seen in children to myasthenic mothers and in patients treated with penicillamine. A relative excess of anti-Id at birth is negatively correlated to the disease neonatal myasthenia. Anti-Id production is found in both healthy and affected children as the concentration of receptor antibodies decreases. The same switch from Id to anti-Id dominance is seen after penicillamine treatment. Thus, both in neonatal myasthenia and after cessation of penicillamine treatment, a relative increase in anti-Id accompanies the gradual disappearance of myasthenic symptoms.

Of special interest is the role of receptor-related antibodies of IgM class. They are found in patients with diseases not accompanied by disturbances of neuromuscular function. Anti-Id antibodies in bone marrow-transplanted patients without myasthenia are often of IgM class, as are receptor antibodies, related Id, and anti-Id in patients with primary biliary cirrhosis and hematological disorders. This is in contrast to the findings in myasthenia gravis, where the great majority of receptor-related antibodies are of IgG class. With the exception of early disease, autoantibodies belonging to IgM class can be detected in low concentrations in less than 10% of patients.

The nature of the epitopes recognized by these antibody Id and anti-Id, respectively, remains to be shown. That idiotypic cross reactivity between IgG and IgM antibodies related to the receptor occurs is evident both from our results from patients with bone marrow transplants and hematological disorders described above, and also from experiments using monoclonal human-human hybridoma IgM antibodies that bind to a variety of autoantigens, including the acetylcholine receptor.[9,45] These IgM antibodies may belong to the species of naturally occurring clones producing multireactive, low-affinity IgM antibodies that have been recently described.[46] Such IgM antibodies should have the intrinsic capacity of binding to many epitopes, and also on autoantigens and autoantibodies. This implies the possibility of a high idiotypic connectivity within the immune system. If the immune reaction is allowed to continue, the IgM response is characteristically followed by a more specific IgG response consisting of high affinity antibodies. Since the capacity to make autoantibodies is universally distributed, the appearance of high affinity IgG antibodies directed against epitopes on autoantigens or autoantibodies could be regarded as a failure of down regulation of the maturing B-cell response to a particular epitope. The natural function of auto-anti-Id has been suggested to be to down regulate an autoantibody response.[47] The pattern of autoantibodies related to the receptor during development of the disease and in patients after bone marrow transplantation substantiates the inverse relationship between IgG receptor antibodies and related Id, on one hand, and anti-Id and IgM receptor antibodies on the other, and provides evidence for down regulation. This priority of the anti-Id response is found both in early myasthenia and in patients who develop myasthenia following bone marrow grafting. Signs of myasthenia may never develop as long as the balance between different antibody species is such that the concentration of anti-Id is sufficiently high. The decrease in anti-Id concomitant with the increase in IgG receptor antibodies as the disease progresses should then indicate a failure of the natural defense systems against the expansion of an autoantibody-producing clone. Another possibility must also be taken into account. Species of antibodies, either anti-Id or Id, may bear epitopes resembling structures on the autoantigen or autoantibodies. The recently described antibodies against acetylcholine in myasthenia gravis must be regarded as anti-Id, bearing an internal image of the transmitter binding region of the receptor,[11] and should, thus, be able to stimulate receptor antibody production. Such antibodies may trigger the expansion of B-cell-producing clones bearing the complementary structure, leading ultimately to a sustained antibody production and disease in a susceptible person. Expansion of autoreactive clones could even emerge as by-products of other immune reactions not primarily directed against the acetylcholine receptor, since the epitope triggering the specific autoantibody response may be found on parallel sets of antibodies having irrelevant binding specificities.

Thus, anti-Id may theoretically have protective effects against disease by down regulation of autoantibody-producing clones, and the reverse, enhancement of autoantibody production probably by molecular mimicry of epitopes on autoantigen or autoantibodies. Which of these mechanisms is operating during the development of myasthenia gravis is not possible to postulate from our results. The consistent priority of the anti-Id response certainly suggests that the expression of anti-Id is critical in the induction phase of the disease.

One important result of our studies is that they clearly indicate the need for distinguishing between the diversity of the immune repertoire and the spectrum of pathogenetic antibodies that actually play a role in the disease process. A select subset of antibodies may bind to epitopes on autoantigens or on complementary antibodies by virtue of their affinity, their isotype or their charge, and case tissue may damage or trigger events in the immune regulatory system, yet many others reactive with the same epitope may be detected in circulation. Our studies of the pattern of Id and anti-Id produced in tissue culture as compared to that found in serum indicate that the spectrum of these autoantibodies is more completely revealed in tissue culture systems. Moreover, there is a better correlation between severity of symptoms and the autoantibody pattern in tissue culture than in serum. The implication is that the pathogenetically relevant antibodies may not be the circulating ones, but the ones already bound to tissue antigens or to complementary antibodies, and that critical information may be lost when studying only the serum autoantibody pattern.

To summarize, the main important findings presented here may be considered circumstantial evidence for a functionally important Id-anti-Id network in myasthenia. The high number of clones secreting anti-Id after EBV transformation of peripheral B lymphocytes indicates that the prerequisites for a potent anti-Id response exist. The pathogenetically important antibody species may not be found in serum, because of rapid elimination by binding to antigen or to complementary antibodies. A role of anti-Id in the immune regulation of the disease is suggested mainly from the pattern of autoantibodies during development and subsidence of the myasthenic process.

REFERENCES

1. **Jerne, N. J.,** Towards a network theory of the immune system, *Ann. Immunol. (Inst. Pasteur)*, 125C, 373, 1974.
2. **Bona, C. A. and Pernis, B.,** Idiotypic networks, in *Fundamental Immunology*, Paul, W. E., Ed., Raven Press, New York, 1984, 577.
3. **Lefvert, A. K.,** Anti-idiotypic antibodies in myasthenia gravis, Abstract, presented at 5th Int. Congr. Neuromuscular Diseases, Marseille, September 7 to 11, 1982.
4. **Dwyer, D. S., Bradley, R. J., Urquart, C. K., and Kearney, J. F.,** Naturally occurring anti-idiotypic antibodies in myasthenia gravis patients, *Nature*, 301, 611, 1983.
5. **Lefvert, A. K.,** A regulatory role for anti-idiotypic antibodies in myasthenia gravis, *Acta Neurol. Scand.*, 69 (Suppl. 98), 204, 1984.
6. **Lefvert, A. K.,** Auto-anti-idiotypic immunity and acetylcholine receptors, in *Concepts in Immunopathology — Immunoregulation*, Vol. 3, Cruse, J. M. and Lewis, R. E., Eds., S. Karger, Basel, 1986, 285.
7. **Lefvert, A. K., Sundén, H., and Holm, G.,** Acetylcholine receptor antibodies and anti-idiotypic antibodies produced in blood lymphocyte cultures from patients with myasthenia gravis, *Scand. J. Immunol.*, 23, 655, 1986.
8. **Lefvert, A. K., Holm G. and Pirskanen, R.,** Auto-anti-idiotypic antibodies in myasthenia gravis, in *Myasthenia Gravis*, in press.
9. **Lefvert, A. K.,** Autoantibodies related to the acetylcholine receptor in myasthenia gravis, in *Proc. 10th Argenteuil Symp.*, Raven Press, New York, in press.
10. **Lefvert, A. K.,** Evidence for the existence of an idiotype-anti-idiotype network in human myasthenia gravis, *Plasma Ther. Transfus. Technol.*, 7, 187, 1986.
11. **Souan, M. L., Geffard, M., Lebrun-Grandie, P., and Orgogozo, J. M.,** Detection of anti-acetylcholine antibodies in myasthenic patients, *Neurosci. Lett.*, 64, 23, 1986.
12. **Abdou, N. I., Wall, H., Lindsley, H. B., Halsey, J. F., and Suzuki, T.,** Network theory in autoimmunity: in vitro suppression of serum anti-DNA antibody binding to DNA by anti-idiotypic antibody in systemic lupus erythematosus, *J. Clin. Invest.*, 67, 1297, 1981.
13. **Shechter. Y., Mason, R., Elias, D., and Cohen, I. R.,** Auto-antibodies to insulin receptor spontaneously as anti-idiotypes in mice immunized with insulin, *Science*, 216, 542, 1982.
14. **Cohen, I. R.,** personal communication, 1985.
15. **Teuber, J. and Helme, K.,** Immunoregulation by anti-idiotypic antibodies during the course of autoimmune disorders of the thyroid, *Immunobiology*, 160, 120, 1981.
16. **Lefvert, A. K. and Bergström, K.,** Acetylcholine receptor antibodies in myasthenia gravis: purification and characterization, *Scand. J. Immunol.*, 8, 525, 1978.
17. **Drachman, D. B.,** The biology of myasthenia gravis, *Annu. Rev. Neurosci.*, 4, 195, 1981.
18. **Bergström, K., Franksson, C., Matell, G., and von Reis, G.,** The effect of thoracic duct lymph drainage in myasthenia gravis, *Eur. Neurol.*, 9, 157, 1973.
19. **Pirskanen, R., Svanborg, E., Sundewall, A. C., and Lefvert, A. K.,** Receptor antibodies and SF-EMG findings in myasthenia gravis patients during remission, in *Myasthenia Gravis*, in press.
20. **Lefvert, A. K. and Osterman, P. O.,** Neonates to myasthenic mothers: a clinical study and an investigation of kinetic and biochemical properties of the acetylcholine receptor antibodies, *Neurology*, 33, 133, 1983.
21. **Lefvert, A. K., Pirskanen, R., and Svanborg, E.,** Anti-idiotypic antibodies, acetylcholine receptor antibodies and disturbed neuromuscular function in healthy relatives to patients with myasthenia gravis, *J. Neuroimmunol.*, 9, 41, 1985.
22. **Lefvert, A. K., Priskanen, R., and Svanborg, E.,** Healthy relatives of patients with myasthenia gravis have low concentrations of acetylcholine receptor antibodies, of immunoglobulins bearing associated idiotypes and of anti-idiotypic antibodies, in *Myasthenia Gravis*, in press.
23. **Sundewall, A. C., Lefvert, A. K., and Olsson, R.,** Anti-acetylcholine receptor antibodies in primary biliary cirrhosis, *Acta Med. Scand.*, 217, 519, 1985.
24. **Sundewall, A. C., Norberg, R., and Lefvert, A. K.,** Cross-reactivity of acetylcholine receptor antibodies and mitochondrial antibodies, in *Myasthenia Gravis*, in press.
25. **Gilhus, N. E., Aarli, J. A., Janzen, R. W. C., Otto, H. F., and Matré, R.,** Skeletal muscle antibodies in thymoma patients without myasthenia, *Acta Neurol. Scand.*, 69 (Suppl. 98), 212, 1984.
26. **Lefvert, A. K. and Björkholm, M.,** Antibodies against the acetylcholine receptor in hematological disorders: implications for the development of myastenia gravis after bone marrow grafting, *N. Engl. J. Med.*, 317, 170, 1987.
27. **Smith, C. I. E., Hammarström, L., and Lefvert, A. K.,** Bone marrow grafting induces acetylcholine receptor antibody formation, *Lancet*, 1, 978, 1985.
28. **Lefvert, A. K., Bolme, P., Lönnquist, B., Ringdén, O., Slördal, S., and Smith, E.,** Bone marrow transplantation selectively increases the production of acetylcholine receptor antibodies, related idiotypes and anti-idiotypic antibodies, in *Myasthenia Gravis*, in press.

29. **Lefvert, A. K., James, R. W., Alliod, C., and Fulpius, B. W.**, A monoclonal anti-idiotypic antibody against anti-receptor antibodies from myasthenic sera, *Eur. J. Immunol.*, 12, 790, 1982.
30. **Lefvert, A. K. and Fulpius, B. W.**, Receptor-like activity of a monoclonal anti-idiotypic antibody against an anti-acetylcholine receptor antibody, *Scand. J. Immunol.*, 19, 485, 1984.
31. **Lefvert, A. K., Holm, G., Sundén, H., and Pirskanen, R.**, Cellular production of antibodies related to the acetylcholine receptor in myasthenia gravis: correlation with clinical stage, *Scand. J. Immunol.*, 25, 265, 1987.
32. **Oosterhuis, H. J. G. H.**, Studies in myasthenia gravis. I. A clinical study of 180 patients, *J. Neurol. Sci.*, 1, 512, 1964.
33. **Smith, C. I. E., Aarli, J. A., Biberfeld, P., Bolme, P., Christensson, B., Gahrton, G., Hammarström, L., Lefvert, A. K., Lönnqvist, B., Matell, G., Pirskanen, R., Ringden, O., and Svanborg, E.**, Myasthenia after bone marrow transplantation: evidence for a donor origin, *N. Engl. J. Med.*, 309, 1565, 1983.
34. **Slördahl, S.**, personal communication, manuscript.
35. **Cerny, J. and Kelsoe, G.**, Priority of the anti-idiotypic response after antigen administration: artefact or intriguing network mechanism?, *Immunol. Today*, 5, 61, 1984.
36. **Namba, T., Brown, S. B., and Grob, D.**, Neonatal myasthenia gravis: report of two cases and review of the literature, *Pediatrics*, 45, 488, 1970.
37. **Vincent, A. and Newsom-Davis, J.**, Anti-acetylcholine receptor antibodies in D-penicillamine-associated myasthenia gravis, *Lancet*, 1, 1254, 1978.
38. **Bolger, G. B., Sullivan, K. M., Spence, A. M., Appelbaum, F. R., Johnston, R., Sanders, J. E., Deeg, H. J., Witherspoon, R. P., Doney, K. C., Nims, J., Thomas, E. D., and Storb, R.**, Myasthenia gravis after allogeneic bone marrow transplantation: relationship to chronic graft versus host disease, *Neurology*, 36, 1087, 1986.
39. **Seely, E., Drachman, D., Smith, Br. R., Antin, J. H., Ginsburg, D., and Rappeport, J. M.**, Post bone marrow transplantation (BMT) myasthenia gravis: evidence for acetylcholine receptor abnormality, *Blood*, 64(Suppl. 1) (Abstr.), 221a, 1984.
40. **Cain, G. R., Cardinet, G. H., III, Cuddon, P. A., Gale, R. P., and Champlin R.**, Myasthenia gravis and polymyositis in a dog following fetal hematopoetic cell transplantation, *Transplantation*, 41, 21, 1986.
41. **Shulman, H. M., Sullivan, K. M., and Weiden, P. L.**, Chronic graft versus host syndrome in man. A long-term clinicopathologic study of 20 Seattle patients, *Am. J. Med.*, 69, 204, 1980.
42. **Weinbaum, J. G. and Thomson, R. F.**, Erythroblastic hypoplasia associated with thymic tumor and myasthenia gravis, *Am. J. Clin. Pathol.*, 25, 761, 1955.
43. **Steinitz, M., Klein, G., Koskimies, S., and Maker, O.**, EB virus induced B lymphocyte cell lines producing specific antibody, *Nature*, 269, 420, 1977.
44. **Stålberg, E., Trontelj, J. V., and Schwartz, M. S.**, Single fibre recordings of the jitter phenomenon in patients with myasthenia gravis and in members of their families, *Ann. N.Y. Acad.Sci.*, 274, 189, 1976.
45. **Duggan, D.B., Mackworth-Young, C., Lefvert, A. K., André-Schwartz, J., Mudd, D., McAdam, K.P.W.J., and Schwartz, R. S.**, Polyspecificity of human monoclonal antibodies reactive with *Mycobacterium leprae*, mitochondria, SS DNA, cytoskeletal proteins and the acetylcholine receptor, in press, 1987.
46. **Holmberg, D., Forsgren, S., Ivars, F., and Coutinho, A.**, Reactions among IgM antibodies derived from normal neonatal mice, *Eur. J. Immunol.*, 14, 435, 1984.
47. **Thorbecke, G. J., Bhogal, B. S., and Siskind, G. W.**, Possible mechanism for down-regulation of autoantibody production by auto-antiidiotype, *Immunol. Today*, 5, 92, 1984.

Chapter 6

ANTI-IDIOTYPE ANTIBODIES MIMICKING ACTIVE BIOLOGICAL SUBSTANCES*

N. R. Farid

TABLE OF CONTENTS

* Original work from the author's laboratory was supported by grants from the Medical Research Council of Canada.

I. INTRODUCTION

An expanded and more comprehensive formulation of Jerne's network theory[1,2] of the immune system requires that it be self-centered, self-referential, and complete.[3,4] In other words, such a system will have a complete library of "internal images" of self- and any potential nonself-antigens. A major activity of the immune system involves the generation of anti-idiotype (anti-Id) antibodies, B cells, T cells, and of various classes. Encounters with foreign antigens incidentally disturb prevailing dynamic equilibrium. A medium-term self-limiting response to a foreign antigen would result in the reestablishment of a new steady state and enrichment of immunological memory. "Holes" in the repertoire of "internal images" could potentially result in the demise of the individual. In an almost heretical view of the evolution of the immune system and the environment, Köhler[5] has suggested that the environment may have evolved to accommodate some of the major holes in the immune system repertoire.

In this context, the "internal image" concept is more than just a laboratory curiosity or a novel means of inducing antigen-specific immunity and occupies a central position in the interconnection between components of the system.

The structural basis of the internal image idiotypes (Id) remains a major enigma in idiotypy. Internal image anti-Id account for a small percentage of an anti-Id response, are known not to be restricted to isotypes or V_H families, and can give rise to seemingly unpredictable varieties of complementary Id.[6] By contrast, the Ab_1' class of anti-anti-Id shows remarkable sequence homology with Idiotypes,[7] as would be anticipated from the workings of regulatory Id.[8] The interaction of the vast majority of "internal image" Id with their complementary Id is inhibited by antigen; some, however, are antigen noninhibitable.[9] Erlanger[9] has suggested that this may be accounted for by the involvement of the reverse face of the immunoglobulin paratope. Internal image Id may mimic antigen because of the correspondence, on the Id and antigen, of critical sequences either in linear arrangement or dispersed along the polypeptide chain, but brought together by protein folding. From a statistical viewpoint, sequence homology is unlikely to account for the molecular mimicry between internal images and antigen. A variety of conformational similarities at contact points of antigen or Id is likely to account for the balance of the mimicry.[10] This, rather than conservation of idiotypic determinants among evolutionarily widely separated phyla, would account for the ability of "internal image" anti-Id to bind with high affinity to Id raised in a variety of species. These subjects are dealt with in detail elsewhere in this monograph.

II. "INTERNAL IMAGE" ANTI-ID AND RECEPTORS

Most studies concerned with Id recognition involving antigens without known biological functions have restricted themselves to the interaction between the idiotypic determinants of the two antibodies and whether or not this was inhibitable by haptens.[11,12] The potential for an "internal image" class of anti-Id antibodies as a means of probing hormone receptors has, however, drawn interest recently.[13,14] We discussed earlier[14] the hypothesis that the ancestral Ig gene may have evolved to give rise, on the one hand, to the Ig superfamily and, on the other, to receptors for these molecules and possibly for those of hormones.[15,16] Attention has furthermore been drawn to the structural and functional similarities between Igs and hormone receptors.[17] Both types of structures are organized into functional domains consisting of a ligand-recognizing component which interacts with a domain able to transmit signals to an effector system. Given the potential similarities between receptor and Igs, it would not be unreasonable to expect that internal image Id, including epibodies,[18] may contribute to the biological activity of anti-Id antibodies.

I discuss in this chapter the use of ligand-specific anti-Id to raise receptor-specific antibodies as a means of investigating the interaction of various ligands with their receptors. I have attempted as much as possible to arrange studies in chronological order so that the reader may gain perspective of the unfolding conceptual progress.

A. Anti-Id Antibodies against Insulin Antibodies

Sege and Petersen have to be credited with demonstrating the feasibility of using antihormone[19] and antivitamin transporting-protein (see below)[20] anti-Id as a means of probing their respective receptors. Antiinsulin anti-Id were prepared by immunizing rats with bovine insulin and, subsequently, immunizing rabbits with affinity-column purified insulin-specific antibody. IgG fractions from the anti-Id antisera were isolated on *Staphylococcus aureus* protein A affinity column; antibodies against isotypic and allotypic rat IgG determinants were removed on rat IgG immunoadsorbent. Before use, the anti-Id IgG was passed over insulin-affinity column to remove any possible antiinsulin antibodies followed by gel filtration to eliminate insulin from the preparation. The anti-Id nature of the IgG was demonstrated by its ability to bind radiolabeled Fab fragment of the original Id (but not that of antiretinol binding protein [RBP]), as demonstrated by gel filtration. Furthermore, insulin, but not RBP, inhibited the immunoprecipitation of Fab of Id by anti-Id IgG.

Anti-Id IgG inhibited the binding of ^{125}I-insulin to epididymal fat cells in a dose-dependent manner (~2 mg/mℓ caused 50% inhibition), whereas normal IgG had no effect, suggesting that the anti-Id interacts with or is adjacent to the insulin receptor on the cell membrane. Likewise, α-aminoisobutyric acid uptake by young rat thymocytes was increased by 5 mg/mℓ of antibody to the same degree as 5 μg insulin per milliliter. This pioneering work was not immediately embraced because of the low bioactivity of the anti-Id, the contention that normal IgG mediated the effects measured,[21] and the inherent lack of a measure of insulin-dependent signal transmission and effector activity.[17]

It was later shown[23] that mice immunized with insulin raise antiinsulin receptor antibodies. The "antireceptor" antibody preparation inhibited ^{125}I-insulin binding to fat cells, stimulated glucose oxidation and lipogenesis, as well as inhibited lipolysis — all insulin-like effects. The putative antireceptor antibodies' interaction with the receptor was blocked by antiinsulin antibodies which also bound a population of the antibodies with antireceptor activity. Mice developing antiinsulin anti-Id manifested fasting hypoglycemia following the first peak of anti-Id, which was then replaced by hyperglycemia following a second peak of antibodies.[23] The presence of these antibodies was associated with reduced insulin receptor numbers (down regulation) and insulin insensitivity to insulin action.[23] The authors considered the possibility that the antireceptor activity may be related to insulin complexed to antiinsulin antibody (which would bind to the receptor) or to auto-anti-Id with an "internal image" of insulin. The anti-Id nature of the antireceptor antibody was suggested by the fact that the antibody exhibited the same activity after the dissociation of any insulin/antiinsulin immune complexes followed by gel chromatography to separate insulin from antibody. Moreover, mild trypsinization, a treatment which greatly reduces insulin responsiveness, was without effect on the activity of the putative antireceptor antibody and was reminiscent of the insulin-like activity of concanavalin A,[23] i.e., on a receptor domain separate from that recognized by insulin. This suggestion, however, contradicts the ability of anti-Id preparation to inhibit receptor binding of ^{125}I-insulin. It was later demonstrated using insulin molecule analogues that whether or not an anti-Id with biological activity at the insulin receptor is obtained depends on whether or not Id were raised with insulin molecule analogues capable of binding to the receptor.[24]

Maron et al.[25] also noted that 45% of patients newly diagnosed with Type I diabetes exhibit antiinsulin receptor antibodies, which made the authors wonder whether these may be antiinsulin anti-Id. This view is particularly intriguing, since 16% of such patients were found to have circulating antiinsulin antibodies at the time of diagnosis[26,27] and two out of five patients who were initially negative developed antiinsulin receptor antibodies after treatment with insulin.[25] Antiinsulin antibodies likely represent part of the autoimmune response to pancreatic β cells and their products as discussed elsewhere.[27]

B. Anti-Id Antibodies Raised against Antibodies to Retinol-Binding Protein

Antiretinol-binding protein (RBP) anti-Id was prepared and tested for specificity by Sege and Petersen,[19,20] as was outlined above for the antiinsulin anti-Id. The biologic activity of the anti-RBP anti-Id IgG was shown by its ability to bind to RBP receptors on the surface of rat intestine epithelial cells, well above the binding by preimmune IgG; this binding was abolished by anti-RBP. Moreover, the anti-Id interfered with RBP-mediated uptake of [³H]-retinol by intestinal epithelial cells. It was not reported whether RBP inhibited the binding of anti-Id to these epithelial cells nor the binding of RBP by anti-Id. Earlier these investigators showed the anti-Id to interact with prealbumin, a physiological carrier for RBP.[20] The binding of the anti-Id to prealbumin was found to be inhibited by anti-RBP antibodies and RBP (but not by soybean trypsin inhibitor), suggesting that the anti-Id bound to the RBP recognition site on prealbumin.

That anti-Id may deliver unequivocal signals characteristic of the ligand antibodies to which they have been raised had to await the demonstration of ligand-specific signal transmission and effector functions consequent upon the binding of anti-Id to the receptor. These requirements were satisfied with anti-Id raised against anti-β-adrenergic antagonist and anti-TSH antibodies.

C. Anti-β-Adrenergic Ligand Anti-Id

Anti-Id were raised to the β-adrenergic antagonist alprenolol by two laboratories,[28,29] instigated in one laboratory[29] by the previous finding of an antiserum fraction in a rabbit immunized with alprenolol of antibodies with β-adrenergic receptor activity,[30] i.e., containing an "internal image" auto-anti-Id. Schreiber et al.[31,32] determined that the antialprenolol Id was specific for the ethanolamine side chain, common to β-adrenergic agonists and antagonists. Anti-Id antibodies raised as a result of immunizing another set of rabbits with purified antialprenolol antibodies were found to behave either as agonists[28] or antagonists.[29]

The differences in subsequent anti-Id activity are likely due to selection of Id as well as the immunization protocol to raise anti-Id. These apparently interacted with different domains of the β-adrenergic receptor. Thus, whereas Schreiber et al.[28] used DEAE column to purify Ig which was injected weekly for 4 weeks into allotype-matched rabbits, Homcy et al.[29] obtained Id for immunization off an acetubtolol affinity resin by L-propranolol elution and injected these antibodies into allotype-matched rabbits every 3 weeks for 6 months. The anti-Id obtained by Homcy et al.[29] was shown to behave as a competitive inhibitor of β-adrenergic agonists in saturation binding analysis and by their effect on basal as well as isoprotrenol-mediated adenylate cyclase activation. The anti-Id of Schreiber et al.[28] behaved as noncompetitive inhibitors as demonstrated by their influence on β-adrenergic receptor number vs. binding affinity, their capacity to enhance β-adrenergic agonist stimulation (but not that of fluoride or Gpp[NH]p) of adenylate cyclase, as well as to enhance basal cyclase activity in their own right.[31] These findings suggested that the antibodies bind at a site on the β-adrenergic receptor different from that of catecholamine agonist or antagonists.[33] Both varieties of anti-Id,[28,29] however, had in common the ability to inhibit binding by [³H]-alprenolol to its Id and the inhibition of agonist and antagonist ligands binding to their specific receptors.

Strosberg and colleagues (reviewed in Reference 33) were able to visualize by indirect immunofluorescence antialprenolol anti-Id binding sites on cells known to have β-adrenergic receptors, but not those without such receptors. Radiolabeled anti-Id binding sites on various tissues correlated closely with receptor number determined by conventional ligands. Lastly,

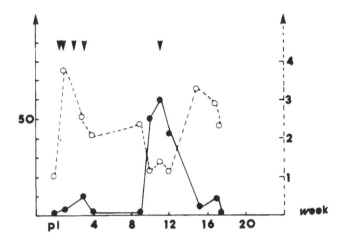

FIGURE 1. The time course of antialprenolol anti-Id and anti-anti-Id activities. Each frame represents a separate rabbit. Anti-Id activity (●——●) is expressed as the inhibition of [³H]-dehydrolalprenolol (DHA) binding to turkey erythrocyte membranes, which are rich in β-adrenergic receptors. The anti-Id was purified from each bleeding by affinity chromatography on an antialprenolol (Id)-sepharose gel. Anti-anti-Id activity (○——○) is expressed as [³H]-DHA binding capacity of the corresponding effluents not retained on the above affinity column. Note should be made of the reciprocal cyclicity of the anti-Id and anti-anti-Id, of the individual variation from one rabbit to another. The animal in the lower part of figure demonstrates only an anti-Id response coincident with the fifth boost (▼). Then the response is completely lost beyond week 15. (Reproduced from *The Journal of Experimental Medicine*, 1983, 157, 1369, by copyright permission of the Rockefeller University Press.)

this group was able to precipitate with antialprenolol anti-Id the 60-kdalton component of the β-adrenergic receptor.[30,32]

Examination of the temporal appearance and decline of anti-β-adrenergic antagonist anti-Id yields valuable lessons[34] concerning the reciprocal cyclicity of Id and anti-Id[35] and the role of regulatory idiotopes.[8] The appearance of antialprenolol anti-Id[34] in response to immunization with the anti-β-adrenergic antibody was transient and occurred in sharp peaks, which differed from rabbit to rabbit (Figure 1). When the alprenolol binding capacity, as well as antireceptor activity of sera taken at different times of the immunization schedule, was examined, it became apparent that peaks of antibodies with catecholamine binding ability (Ab$_1$.) alternated with peaks of receptor-binding anti-Id (Ab$_2$) (Figure 1). In accordance with precepts of the Id-anti-Id interactions,[1,2,8] it could be demonstrated that after the separation of the different antibody subpopulations on the appropriate affinity columns the Id neutralizes the anti-Id and the anti-anti-Id neutralizes the anti-Id. The anti-Id preparation was without [³H]-alprenolol binding capacity, whereas the anti-anti-Id had no activity at the β-adrenergic receptor.

It has, however, to be emphasized that as the immune response proceeds down the cascade of complementary antibodies, a drift from the initial stimulus is observed. Thus, whereas the antialprenolol Id and the anti-anti-Id bound [³H]-alprenolol, they did so with different affinities. Initially, the anti-anti-Id bound [³H]-alprenolol with a higher affinity than the Id! In successive weeks, however, the affinity of the anti-anti-Id decreased progressively below the affinity of the Id,[34] i.e., suggesting that more and more of the anti-anti-Id were directed against IgV framework determinants and may well be regulatory in nature[8]

Strosberg's laboratory has extended its observation on Id mimicking β-adrenergic agonists to monoclonals.[35] Hybridoma cells making monoclonals against alprenolol were used as immunogen to raise monoclonal anti-Id. Six antibodies inhibited [³H]-alprenolol binding to Id and three of these recognized β-adrenergic receptors. One anti-Id monoclonal, studied in detail, bound to cellular β-adrenergic receptors, stimulated adenylate cyclase, and precipitated a single 55-dalton peptide.

Of particular and general interest is the finding by the laboratory of Greene that the receptor of reovirus Type 3 isolated by antiviral anti-Id represents the β-adrenergic receptor[37,38] (also, see below).

D. Antiglycoprotein Hormone Anti-Id
1. Antithyrotropin Anti-Id

We raised "internal image" anti-TSH anti-Id with the intention of answering the following questions. Would such antibodies bind to TSH receptor? If they did bind, would they behave as agonists or antagonists? If the anti-Id are agonists, would they sustain the medium- and long-term biologic effects of TSH thought to be, at least in part, mediated through cAMP? And lastly, could such anti-Id account for all or most of the attributes of Graves' IgG antibodies? These antibodies are causally related to the hyperthyroidism of Graves' disease (reviewed in Reference 11). At the time of the initiation of the study there was no consensus (and no means of finding out) whether these antibodies were directed against the receptor or against adjacent components of the plasma membrane (reviewed in References 11 and 39).

Sprague-Dawley rats were immunized with hTSH or bTSH and specific antibodies eluted off a TSH affinity column and were used to immunize rabbits.[40] After absorption with pooled Sprague-Dawley Ig, rabbit antibodies were found to quantitatively inhibit ^{125}I-bTSH binding to the Id.

The anti-Id effected a dose-dependent inhibition of ^{125}I TSH binding to thyroid plasma membrane. The dose-response kinetics of this inhibition yielded a curvilinear relationship; the high affinity binding site of the antibody was assigned a KD of $7.6 \times 10^{-9}\,M$ compared to $1.3 \times 10^{-10}\,M$ for bTSH, suggesting that a species of IgG ($\sim 3\%$) in the anti-Id preparation interacted with the TSH receptor. The binding of radiolabeled anti-Id to thyroid plasma membrane was saturable and some two thirds of the ^{125}I-anti-Id bound to receptor was inhibited by 160 mU of TSH per milliliter, emphasizing the high affinity binding of these antibodies to the TSH receptor. Moreover, the anti-Id induced the activation of the thyroid plasma membrane adenylate cyclase when tested at 37°C in presence of Gpp(NH)p. Interestingly, the anti-Id antibody preparation inhibited adenylate cyclase to a degree greater than normal rabbit IgG in the absence of Gpp(NH)p at 30°C. Considering that the anti-TSH anti-Id is a polyclonal antibody preparation which contains a mixture of agonists and antagonists, it was suggested that the action of the antagonist was favored under these experimental conditions (also, see below). Lastly, these antibodies induced two TSH-specific functions mediated by cAMP: the uptake iodide into isolated thyroid cells and the organization of these cells into follicles. Interestingly, the anti-Id simulated the well-described bimodal effect of TSH on iodide transport[41,42] in that net exit of ^{131}I from thyroid cells was noted after 10 min incubation and a significant increase above control after 4 hr incubation with 100 μg/ mℓ anti-Id IgG. The anti-Id raised against antibody specific for intact TSH thus simulated all the actions of TSH investigated. It presented a tool whereby to probe the TSH receptor.

It was apparent, moreover, that an antibody (irrespective of the manner in which it was raised) specific for the TSH receptor possessed all the attributes of Graves' IgG. We could not, however, exclude the possibility that the reputedly polyclonal Graves' IgG may contain Igs directed against membrane structures other than the receptor and which may contribute to variation in the biologic activity of different Graves' Igs. On this basis, we suggested that Graves' IgG may be an auto anti-Id, at least in some instances.[11,40] This hypothesis would predict that at least some individuals (possibly at increased risk of developing Graves' disease) would have antibodies capable of specifically binding TSH. This, indeed, has recently been found to be the case.[43,44] These "natural antibodies" are presumably characterized by a high degree of idiotypic connectivity,[45] i.e., under stringent control by other members of the same set. Disturbance of these checks could potentially lead to appearance

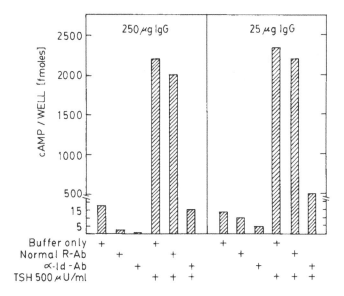

FIGURE 2. The influence of TSH anti-anti-Id on basal and bTSH-driven cyclase activity. The serum was obtained from the same rabbit 2 weeks after the bleed resulting in the anti-Id IgG which mimicked TSH described above. The ability of serum in presence of bTSH to stimulate cAMP generation in primary culture of human thyroid cells (5×10^4) over 2 hr was investigated. As little as 25 μg/mℓ anti-Id IgG inhibited basal and even remarkable TSH-driven cAMP generation. This inhibitory effect was attributed to the appearance of auto-anti-Id in the animal immunized with rat anti-TSH Id. (From Farid, N. R., et al., *Can. J. Biochem. Cell Biol.*, 62, 1255, 1984. With permission.)

of "internal image" anti-TSH receptor antibodies. As an alternative explanation, ^{125}I-TSH binding antibodies[11] may be anti-Graves' IgG anti-Idiotypes,[44] although this notion cannot explain their presence in a high proportion of healthy persons.

Later sampling of a rabbit which initially produced an antibody with the characteristics described above, yielded IgG which, instead, inhibited basal cyclase activity and almost completely blocked TSH-driven cyclase activation (Figure 2). The predominance of anti-anti-Id (Ab_3) at the time of the later sampling could account for these observations.

The porcine TSH receptor is an $M_r \sim 200,000$ dimeric glycoprotein[46] (see Reference 11 for review). Recent investigations in this laboratory suggest the porcine TSH receptor subunit to be an $M_r \sim 92,000$ structure which is proteolytically degraded to the smaller fragments (M_r 66,000 and 35,000) by plasma membrane-associated proteases triggered by TSH binding.[89] TSH binds not only to the holoreceptor ($M_r \sim 200,000$), but also to breakdown products of the receptor, although with lower affinity.

The anti-TSH anti-Id interacted with an M_r 197,000 band resolved from thyroid plasma membrane under nonreducing conditions and transferred to nitrocellulose paper. This is the same band with which TSH, but not HCG or insulin, interacts (Figure 3). Moreover, prior incubation of the protein blot with TSH blocked the binding of the anti-Id to M_r 197,000.[47] Under these experimental conditions neither TSH nor anti-Id interacted with membrane polypeptides of lower molecular weight resolved under nonreducing conditions, and no binding occurred with membrane polypeptides resolved under reducing conditions. Anti-TSH anti-Id was also found to preferentially inhibit photoaffinity cross-linking of ^{125}I-bTSH to the holoreceptor when membranes were resolved under nonreducing conditions.[48] We, therefore, concluded that the "internal image"-type antibodies in the anti-TSH anti-Id recognized the receptor domains which make contact with TSH.

2. Anti-TSH Subunit Anti-Id Antibodies

We next turned our attention to the mode of interaction of TSH with its receptor.[49] TSH

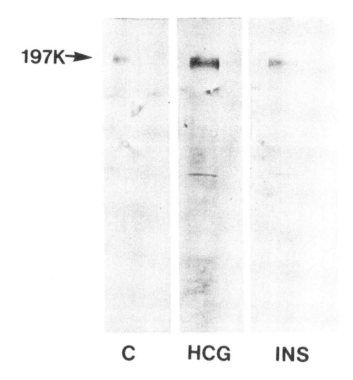

197K→

C   HCG   INS

FIGURE 3.   Binding of anti-TSH anti-Id to the TSH holoreceptor (M_r = 197kdaltons). Thyroid plasma membranes were resolved on 7.5 to 15% SDS-polyacrylamide gradient gel. The peptides resolved were protein blotted and allowed to interact with anti-TSH anti-Id IgG. This antibody interacted with the M_r- 197,000-dalton TSH holo-receptor. Preincubation of the paper strips with 500 IU of human gonadotropin (HCG) or insulin (INS) did not influence the binding of the anti-Id to the receptor band. TSH, by contrast, inhibited markedly the binding of the anti-Id to M_r 197,000 band. (From Farid, N. R., et al., *Can. J. Biochem. Cell. Biol.*, 62, 1255, 1984. With permission.)

shares with FSH, LH, and hCG a common α subunit which interacts noncovalently with a hormone-specific β subunit.[50] The deduction that β subunit determines hormonal specificity has been adequately substantiated.[50] The way in which the α and β subunits of TSH interact with the receptor to produce hormone-specific signals was not, however, amenable to investigation, because the free dissociated subunits are without biological effect. A number of models for the TSH receptor interaction may be envisaged. Thus, it is possible that the β subunit binds to a stereospecific site and the α subunit delivers the hormone signal, or, conversely, the α subunit may bind to a site common for glycoprotein hormone receptor and then β subunit deliver the TSH-specific signal. TSH interaction with receptor may be related to two signals delivered by the two subunits or to a one-combinatorial signal.

In order to resolve this question, we have raised in rabbits, anti-Id antibodies against rat monoclonals, respectively, specific for the α and β subunits of hTSH.[49] We were also curious to learn whether two "internal image" anti-Id could cooperate to simulate the action of TSH at its receptor. The characteristics of the monoclonal Id are outlined in Table 1.

The binding of ^{125}I-hTSH to anti-β subunit monoclonal was inhibited by its complementary anti-Id in a dose-dependent manner; up to 48% at 500 μg IgG per milliliter. By contrast, ^{125}I-hTSH interaction with the anti-α subunit monoclonal was not influenced by up to 1000 μg IgG per milliliter anti-α anti-Id. In a strict sense, it would have been expected that only anti-β anti-Id would include "internal image" molecules. In fact, a specific combination of anti-β anti-Id and one anti-α anti-Id displayed TSH-like activity; preimmune IgG and two other IgG raised to α subunit monoclonal could not be substituted. The anti-α anti-Id

Table 1
CHARACTERISTICS OF ANTI-TSH SUBUNIT-SPECIFIC MONOCLONALS

	Anti-α	Anti-β
Species	Rat	Rat
Ig isotype	IgM	IgG
Binding to ^{125}I-hTSH	+	+
Binding affinity (K_D)	$5.9\ M^{-9}$	$3.7\ M^{-10}$
Displacement of bound ^{125}I-hTSH by bTSH	<5%	<3%
Glycoprotein hormone causing displacement of bound ^{125}I-hTSH	hTSH, hFSH, hLH, hCH	hTSH
Inhibition of membrane bounding ^{125}I-bTSH	−	−
Inhibition of adenylate cyclase activation	+	+ +
Effect on bTSH-induced activation of cyclase		
Added simultaneously	Inhibition	Inhibition
Added 30 min after bTSH	Stimulation	Stimulation

Table 2
ADENYLATE CYCLASE ACTIVATION BY ANTI-α AND ANTI-β TSH SUBUNIT ANTI-ID

	Normal rabbit IgG	Anti-Id			
		Anti-α[a]	Anti-β[a]	Anti-α[a] + anti-β[a]	TSH[b]
Percent cyclase stimulation relative to buffer	− 57 ± 2	− 7.3 ± 3.1	− 8.1 ± 0.9	87.3 ± 4.6	93 ± 6.8

[a] IgG was tested at a concentration of 250 mg/mℓ; in the anti-Id mixture, 125 mg/mℓ of each was used. NRIgG and individual anti-Id are inhibitory.

[b] 250 mU/mℓ crude bTSH (USV Laboratories, Mississauga, Ontario) was used.

in question may be complementary to idiotopes in close proximity to the antigen binding site or may represent a variant of "internal image" anti-Id not inhibitable by antigen.[9]

The anti-Id preparations were without effect each alone or in combination when tested for their ability to inhibit ^{125}I-bTSH binding to thyroid plasma membranes, in spite of the ability of iodinated anti-Id to show binding to thyroid plasma membranes. The binding of equimolar concentrations of the radiolabeled anti-Id was additive. Some 40 to 65% of the bound anti-α anti-β anti-Id and their mixture was inhibited by 500 mU/mℓ of bTSH. The residual binding in the case of the anti-α anti-Id, and particularly with the combination of anti-α and β anti-Id, was well above that for normal rabbit IgG and was attributed to the stabilization of the receptor by these antibodies.[49]

Anti-α and anti-β anti-Id each had some stimulatory activity on basal thyroid membrane adenylate cyclase compared to normal rabbit IgG. Equimolar mixture of the two anti-Id resulted in a marked stimulation of adenylate cyclase comparable to that produced by 150 mU/mℓ of bTSH (Table 2). The more than additive influence of the two anti-Id suggests cooperative interaction, i.e., the binding of one antibody to the TSH receptor increased the affinity of binding of the second antibody.[51] The combination of anti-α and anti-β anti-Id was found to induce dispersed thyroid to organize into follicular structures and to promote the uptake of ^{131}I by these cells. The anti-β anti-Id alone showed some effect in both assays and anti-α was without influence on either.

Lastly, the capacity of these antibodies to bind to protein blots of thyroid membranes resolved under nonreducing conditions was tested. Each of the two anti-Id tested alone did not interact with the blotted thyroid membrane peptides. Equimolar quantities of the two anti-Id consistently yielded a clear band at $M_r \sim 197,000$ daltons. The seeming paradox between the binding of individual radiolabeled anti-Id to thyroid membranes and the inability to demonstrate binding to protein-blotted receptor can be resolved by noting that protein blots are extensively washed in 200 mM NaCl to minimize nonspecific binding. This procedure removes antibodies of low binding affinity and again emphasizes the cooperative interaction of the two anti-Idiotypes in binding to the TSH receptor.

Apparently, following binding to its receptor, TSH induces microaggregation of the receptor which is necessary for triggering adenylate cyclase activation.[31] While it is conceivable that the combination of the two anti-Id act through promoting receptor aggregation, we would need to invoke, in addition, specific interaction of each anti-Id with separate domains of the receptor and the bivalency of each of the two antibodies to account for the slight biologic influence of anti-β anti-Id and the cooperative influence of the two anti-Id, particularly on protein-blotted receptor. The lack of an "internal image" anti-α anti-Id does not allow rigorous testing of the question of specificity.

This study represents the first example of the use of two anti-Id to explore their interaction in mediating hormone-specific signals. These results are all the more intriguing because only one of the two anti-Id is an "internal image" anti-Id in the classic sense.

We were able to reach a number of conclusions.[49] The α and β subunits of TSH deliver two cooperative signals to the TSH receptor. The hormone specificity is, indeed, associated with the β subunit, the α subunit's contribution being through increasing the affinity of the receptor for subsequent β subunit interaction. It has, however, not been possible for us to find out the order in which the two subunits interact with the receptor, as has been determined for HCG.[52] This cooperative interaction is apparently crucial for a receptor conformation requisite for the stimulation of the cyclase system. Studies comparing the interaction of free and receptor-bound bTSH with the monoclonal Id also suggest that after binding to the receptor the bTSH molecule undergoes conformational change(s) such that new antigenic sites previously unrecognized by these monoclonals are exposed (Table 1). The TSH receptor apparently comprises at least two functional domains, one of which is concerned with ^{125}I-bTSH binding and the second with stimulation of the cyclase system, a view consistent with work utilizing anti-TSH receptor monoclonal antibodies as well as a variety of mutually occurring autoantibodies.[47,48,53,54]

We further demonstrated the utility of the anti-TSH subunit anti-Id by using their mixture to identify the TSH receptor and its subunits on protein blots of [3H]-leucine-labeled whole thyroid cell lysates.[55] Following identification, the bands were carefully cut and counted and the results used to monitor biosynthesis and turnover of the TSH receptor. TSH was found to accelerate the biosynthesis of its receptor with very efficient incorporation of receptor subunits into the holoreceptor ($M_r \sim 200$ kdaltons). The effect of TSH upon new receptor synthesis was bimodal both in terms of time and the TSH dose: beyond optimal receptor-labeling times, a progressive decrement in the labeling rate was observed whether or not thyroid cells were grown in the presence of TSH. Likewise, TSH dosage above an optimum of 100 mU/mℓ caused decrement of the synthesis rate of its TSH receptor. To measurably influence receptor biosynthesis, TSH had to be in thyroid cell culture for several hours (<3 and >12), suggesting that TSH effect on receptor synthesis is mediated through transcription and/or mRNA stabilization. TSH was also found to reduce the half-life of the receptor. The most dramatic effect of TSH on total thyroid cell receptor half-life occurred between 12 and 15 hr of incubation. TSH receptor half-life decreased from 7.3 to 4 hr. Interestingly, this coincided with the replacement of the high affinity, low capacity TSH receptors on thyroid cell plasma membranes and their replacement by low affinity, high capacity ^{125}I-bTSH binding sites. This last observation points to a separate, longer-range mechanism of TSH-induced densensitization other than that through the cyclase,[55] which is well documented.

The precepts of the network were recently further explored in immunizing rabbits against purified Graves' IgG either by absorption on guinea pig fat membranes[56] or on affinity column of semipurified porcine TSH receptor.[90] Baker et al.[56] isolated "internal image" anti-Id by absorption on TSH-affinity column; this preparation inhibited binding of Graves' IgG to receptor. Our absorbed rabbit IgG preparation included "internal image" anti-Id, anti-anti-Id (Ab$_3$), as well as an anti-Id, specific for Graves' disease and shared by 10/11 patients tested, directed against idiotopes close to, but distinct from, those to which "internal image" anti-Id bind. The interaction of this Graves' disease cross-reactive Id with the anti-Id apparently interferes with the binding of the "internal image" variety to Graves', but not to Hashimotos or normal IgG. Ab$'_1$-type anti-Id preparation showed dose-dependent stimulation of cyclase both in cultured human and cloned rat thyroid cells, although it did not interfere with the binding of ^{125}I-bTSH to the receptor.

3. Anti-TSH Auto-Anti-Id

Recently, Beall et al.,[57] in an attempt to demonstrate anti-TSH auto-anti-Id production, immunized groups of six rabbits each with crude bTSH (Thyropar), hTSH, β subunit of hTSH, and human serum albumin (as control). IgG from two out of six rabbits in the groups immunized with hTSH, hTSH β,and two out of four surviving bTSH-immunized rabbits were found to bind to human thyroid membranes and to induce adenylate cyclase in cultured thyroid cells. These activities were maximal by 100 days after the sixty booster and subsequently declined gradually. While the inference was that IgG with this activity represents anti-TSH auto-anti-Id, the authors did not demonstrate that the antibodies appearing later did bind to the initially raised anti-TSH or anti-β subunit antibodies nor that this binding is specifically inhibited by the relevant antigens. The authors subsequently found that Ig sera of bTSH-immunized rabbits could bind to an anti-hTSH monoclonal. The results, nevertheless, demonstrate that putative anti-TSH auto-anti-Id show the characteristics of anti-TSH receptor antibodies.

E. Antihuman Chorionic Gonadotropin Anti-Id

Attempts to raise anti-Id antibodies, even against monoclonal antibodies of known narrow specificities, however, may not always be successful. In order to further understand the function of epitopes on HCG molecules, Moyle[91] raised anti-idiotopes against monoclonals specific for α or β subunit determinants,[58] including some which contact the receptor. Rabbits were immunized with monoclonals alone or initially by monoclonals, followed by HCG receptor-rich stimulated granulose cells and monoclonals (hoping to induce receptor-directed anti-Id.) Although the anti-Id bound to monoclonal Id and this interaction was inhibited to ~30% by HCG, the anti-Id were without agonist or antagonist activity at the receptor singly or in combinations.

III. ANTI-ID TO POLYPEPTIDE HORMONE ANTIBODIES OTHER THAN INSULIN

Anti-Id antibodies capable of recognizing the prolactin (PRL) receptor were raised against antibodies to ovine PRL (oPRL) and rat PRL (rPRL).[59] Anti-oPRL or anti-rPRL antibodies were induced in rats or rabbits, respectively, and IgG isolated by protein A-affinity columns. Specific anti-oPRL antibodies were purified on oPRL affinity column. This material and Ig fraction of anti-rPRL were used to immunize rabbits; immune IgG was passed over oPRL affinity column to remove antibodies that might bind labeled oPRL (used in receptor assays).

Binding to PRL receptor was determined both by ^{125}I-protein A or by testing the ability of residual antibody in the supernatant, after precipitating receptor-antibody complex, to inhibit ^{125}I-oPRL binding to anti-oPRL IgG. The membrane binding by anti-Id was competitively inhibited by the immunizing Id. PRL receptor-rich tissue membranes showed fivefold greater binding than heat-inactivated membranes or membranes from tissue with little PRL receptor concentration. Moreover, dose-dependent inhibition of ^{125}I-oPRL binding to its receptor in rat and rabbit PRL receptor-rich membranes was found in anti-rPRL anti-Id IgG. In an earlier study, this group reported an anti-Id raised against rat anti-oPRL, interacted non-competitively with PRL receptor; this antibody was shown to increase the synthesis and secretion of basal and PRL-stimulated α-lactalbumin compared to nonimmune IgG.[60]

IV. ANTI-ID ANTIBODIES INDUCE EXPERIMENTAL MYASTHENIA GRAVIS

The exquisite stereospecificity of anti-Id was demonstrated by Wassermann et al.[61] who used antibodies against *trans*-3-3′-Bis[α-trimethylammonio methyl] azobenzene biomide (BisQ), a potent ligand for the active state of the acetylcholine receptor. BisQ has a molecular weight of 486. A hierarchy of [³H]BisQ binding to Id was found, most prominent for potent agonist and least for potent antagonists. Immunization of rabbits with these Id yielded antisera which interacted with rat, *Torpedo california,* or *Electrophorus electricus* acetylcholine receptors demonstrated by complement fixation and enzyme-linked radioassay; these inter-actions were inhibited by BisQ. Two of the three rabbits immunized showed muscle weak-ness, which was temporarily reversed in one animal injected with neostigmine. Sciatic nerve stimulation showed posttetanic exhaustion of hind limb muscles in another animal. Appar-ently, not all antibodies were equipotent, as some rabbits with high titers of anti-Id antibodies showed no signs of muscle weakness. The anti-Id to BisQ was transient in one animal presented, and with the fall in antibody titer myasthenic muscle weakness improved. Al-though maximal anti-Id titer and its neuromuscular effect occurred after the first boost, the titer started falling after the second boost and fell to undetectable level despite subsequent boosts.

Pursuing the paradigm, Cleveland et al.[62] immunized BALB/c mice with BisQ-bovine serum albumin conjugate and after fusion with myeloma cells selected hybridomas with the ability to bind to anti-BisQ and *Torpedo* acetylcholine receptor, i.e., individual B lympho-cytes making auto-anti-Id were selected for fusion. BisQ inhibited (50% by $4 \times 10^{-5} M$) the binding of the anti-Id monoclonal F8-D5 to the acetylcholine receptor. In a reciprocal experiment, 0.9 μg of acetylcholine receptor gave 50% inhibition. Monoclonal F8-D5 was shown in the presence of 1 mM carbamylcholine, to block by 80% the influx of [134]Cs into a vesicle system containing *Torpedo* acetylcholine receptor, to bind, as detected by indirect immunofluorescence, in a pattern similar to that obtained with highly specific antireceptor antibody, and to adsorb *Torpedo* acetylcholine receptor when immobilized in an affinity column.[63] Interestingly, BALB/c CR mice bearing ascites tumor of F8-D5 hybridoma cells showed different degrees of muscle weakness up to quadriplegia, which is promptly, but temporarily, reversed by injections of neostigmine.[63]

Cleveland et al.[63] explored the wider applicability of the auto-anti-idiotypic monoclonal route for raising receptor antibodies by applying this approach to raising adenosine receptor monoclonals.[63]

V. ANTI-ID ANTIBODIES AS PROBES FOR OPIOID RECEPTORS

Ng and Isom[64] raised, in guinea pigs, antisera against affinity-purified rabbit antibodies specific for 3-O carboxymethylmorphine-BSA. The unfractionated guinea pig sera inhibited the binding of nalaxone to mouse brain homogenates and the contraction of electrically stimulated mouse vas deferens, i.e., interacted with opiate receptors. The authors, however, provided no proof for their contention that the biologic activity in the antisera was related to the presence of anti-Id.

More recently, Glasel and Myers[65] used the F(ab) of a high-affinity murine morphine-specific monoclonal to immunize rabbits. The anti-Id antibody preparation preabsorbed by normal mouse IgG was purified on monoclonal (Id) affinity column. Physiochemical studies of the anti-Id eluted off the column were consistent with a "restricted" polyclonal response. The anti-Id (40 μg/mℓ) affected 80% inhibition of [³H]-morphine binding to the monoclonal Id; normal rabbit IgG was without effect. Anti-Id (5.7 μg/mℓ) caused the virtual abolishment of [³H]-morphine binding to opiate receptor-rich material. Lesser effects were seen with nalaxone and D-Ala-2-D-Leu-5-enkephalin which is not recognized by monoclonal Id. Thus, the anti-Id was found to bind not only to the μ, but also to δ-type opiate receptor, leading the authors to raise questions about a possible close spatial relationship of these two receptor types.

The use of "internal image" approach was also extended to the β-endorphin receptor. Monoclonal 3-E7, which recognizes virtually all known opioid peptides[65] and as such almost certainly interacts with the biologically relevant domain of the opioid receptor,[66] was used to immunize rabbit.[67] Anti-idiotypic sera were isolated by adsorption to an 3-E7 immunoaffinity column followed by elution with the pentapeptide leu-enkephalin which encodes the protein sequence inhibiting 3-E7 monoclonal binding to the opiate receptor. The IgG fraction inhibited β-endorphin binding to 3-E7. The anti-Id inhibited quantitatively the binding of [³H]-diprenorphine to solubilized rat brain opioid receptor, but not to brain membrane fragments; 20 $\mu\ell$ anti-Id, however, inhibited 90% of the binding of ¹²⁵I-β-endorphin membranes. In order to accommodate the fact that 3E7 monoclonal does not bind to diprenorphine, it was suggested that diprenorphine and β-endorphin interact with different sites on the opioid receptor. The antibody caused a dose-dependent inhibition of cAMP synthesis in NG 108CC15 hybrid cells which carry both δ and, possibly, ϵ-type opioid receptor, and this effect was reversed by naloxone; cyclase inhibition observed with 10 $\mu\ell$ anti-Id was equivalent to that with $10^{-5}\,M$ (D-Ala, D-Leu) enkephalin. The authors attributed the lack of anti-Id effect on guinea pig ileum and mouse vas deferens to its inability to reach receptors.

VI. ANTI-ID AS PROBES FOR MEMBRANE TRANSPORT SYSTEMS

A. Glucose Transport

Anti-Id have also been used as a sensitive means for detecting specific membrane transport components, an example in point being the glucose transport component. Until the recent use of cytochalasin B as a photoaffinity label,[68] there was no specific way to identify the glucose transporter. Duronio and Lo[69] recently immunized rabbits with glucosamine-bovine serum albumen conjugates. After purification with protein A-Sepharose® and with glucosamine-Sepharose® 4B, the antiglucosamine IgG (Ab₁) was used to immunize rabbits and a goat. The anti-Id raised in goat were found to interact with two polypeptide bands of M$_r$ 94,000 and 52,000 resolved from L6 rat myoblast membranes; one rabbit anti-Id bound predominantly to a 52 kdalton band, whereas another interacted exclusively with a 94 kdalton protein. Difficulties were encountered in obtaining a good anti-Id response with many rabbits; consequently, the number and type of experiments that could be carried out were limited. The strongest indication of the involvement of the 94K and 52K proteins in glucose transport came from analysis of several glucose transport mutants. Neither protein could be detected with the anti-Id in these mutants. These observations accord with recent identification of Na⁺-D-glucose cotransport protein from the porcine renal proximal tubules,[70] and with the notion that the 94 kdalton peptide is a precursor of the 52K peptide. Kay et al.[71] used a similar anti-Id approach to raise antiglucose transporter and demonstrated that the antibody interacts on protein blots at a zone characteristic of band 3 of RBC membranes. Anti-Id affinity-column purified material was similar to band 3 by peptide mapping. This is in agreement with the suggestions that the RBC glucose transporter (identified as a 50 kdalton polypeptide) may be a proteolytic fragment of band 3.[72]

B. Dicarboxylic Acid Transport Components in *Escherichia coli*

The cell envelope of the Gram-negative bacteria, *E. coli*, is comprised of an outer membrane and a cytoplasmic membrane; the region between these two membranes is referred to as the periplasmic space.[73] At least four components are involved in the translocation of dicarboxylic acids across various regions of the cell envelope. The dicarboxylate binding protein (DBP) and the outer membrane matrix protein are responsible for the outer membrane transport process; whereas DBP by itself facilitates the passage across the highly negatively charged periplasmic space.[73,74] The cytoplasmic membrane dicarboxylate transport system is an active transport process and is dependent on two transmembrane integral proteins, SBP 1 and SBP 2.[73] With the exception of the outer membrane matrix protein, the dicarboxylate transport components are able to bind specifically with the transport substrates, succinate, fumarate, and malate. The binding affinities of these proteins range from 2 to 40 μM for succinate. All three of these components have been purified through the use of aspartate-coupled Sepharose®; aspartate was coupled through its amino group to the Sepharose® backbone through a spacer.[75] The relatively poor substrate binding affinity of the transport components renders it difficult to obtain large quantities of purified transport components.

Walker and Lo[14,92] have recently employed anti-Id antibodies raised in rabbits in conjunction with SDS polyacrylamide gel electrophoresis and Western blot techniques to identify the cytoplasmic membrane dicarboxylate transport components. Antibodies were first raised against aspartate-BSA conjugate. The antiaspartate IgG were affinity purified with an aspartate-Sepharose® column after initial passage of the serum through protein A-Sepharose® and BSA-Sepharose® to select for IgG and remove anti-BSA IgG, respectively. This rabbit antiaspartate (RAA) IgG preparation was then used to raise anti-Id antibodies (rabbit anti-rabbit antiaspartate, RARAA) in a second rabbit. The "internal image" anti-Id would be expected to bind the substrate recognition site on the transport components. Indeed, RARAA IgG from seven rabbits were found to bind with four cytoplasmic membrane proteins; the molecular weights of these proteins are 71, 56, 29, and 20 kdaltons. These anti-Id could also bind with the transport components purified by aspartate-Sepharose® columns. The nature of the above-mentioned four proteins is currently being investigated through the use of appropriate transport mutants.

VII. ANTI-ID AS PROBES FOR FORMYL PEPTIDE RECEPTOR OF THE NEUTROPHIL

Marasco and Becker[76] raised in mice, guinea pigs, and goats anti-Id against rabbit antibodies to the chemoattractant peptide formyl methionyl-leucyl-phenylalanine (formyl Met-Leu-Phe) produced by *Escherichia coli*. Goat antibodies showed the highest titer and were used in all subsequent studies. The goat antibodies bound to rabbit neutrophils and quantitatively inhibited the binding of radiolabeled chemoattractant to these cells; receptor number and specificity identified by labeled anti-Id were comparable to those determined by ligands. Conservation of idiotypic determinants on nearly all rabbit and rat antiformyl Met-Leu-Phe antibodies was verified by the ability of these antibodies to bind (Fab')$_2$ goat anti-Id. Because preimmune IgGs themselves induce chemotaxis and, in the presence or absence of cytocholasin B, granule enzyme release, it was not possible to demonstrate whether anti-Id mimicked the biologic activity of formyl Met-Leu-Phe peptide.[76] Marasco et al.[77] have recently extended these studies by raising in rabbits anti-Id by immunizing with rabbit formyl (f) Met-Leu-Phe with similar characteristics to the earlier antibodies. Interestingly, the rank order of binding of over 40 oligoformylpeptides to rabbit Id correlated closely with their reactivity with neutrophil formyl receptor. In addition, the investigators showed a rapid and cyclic appearance and disappearance of anti-Id; this appears to be mediated by an auto-anti-anti-Id ($Ab_3/Ab_{1'}$) which shared binding characteristics of the Id immunogen (Ab_1).[74] $Ab_{1'}$ had lower binding affinities than Ab_1, for f Met-Leu [^{3}H] Phe was lower than that for Ab_1. The authors excluded the possibility that the ability of anti-Id IgG preparation to inhibit labeled f Met-Leu-Phe binding to the neutrophil receptor was due to ligand binding to Ab_1 by quantitating anti-Id IgG (compared to preimmune IgG) to neutrophil with ^{125}I-protein A.

VIII. THE REOVIRUS TYPE 3 BINDS TO THE B-ADRENERGIC RECEPTOR

One of the most exciting chapters in "internal image" idiotypy has culminated in the discovery that reovirus Type 3 hemagglutinin attachment site is the β-adrenergic receptor.[37,38] The hemagglutinin proteins determine cellular and neural tropism of reovirus. Initially, rabbit anti-Id were raised against a monoclonal (G5) specific for reovirus 3 agglutinin protein and were found to bind to cell types with reovirus Type 3 (but not other type) binding sites, including neurones and T lymphocytes (see Reference 78). An IgM monoclonal anti-Id was subsequently raised by immunizing mice with irradiated G5 hybridoma cells and found to bind only to cells which have reovirus 3 receptor similar to the results of rabbit anti-Id. Reovirus 3-specific cytotoxic T cells lysed in an H2-restricted fashion with hybridoma giving rise to the anti-Id; cytotoxicity was inhibited by G5 Id and was not observed with Sendai virus-specific cytotoxic cells.[79] These findings suggested that reovirus 3-specific cytotoxic T cells must have an antigen receptor capable of recognizing anti-idiotypic determinants on hybridoma cell surfaces. The monoclonal anti-Id was shown to block viral attachment to susceptible cell types, to inhibit viral infectivity, and to generate cell-mediated immunity, as well as to recognize viral receptor.[80-82] The reovirus Type 3 (and anti-Id) binding site was found to correspond to the $M_r \sim 67,000$ β-adrenergic receptor.[37,38] The anti-Id monoclonal precipitated a single monomeric 67,000 peptide from lysates of surface-labeled somatic cells. It precipitated a complex of ^{125}I-iodohydroxybenzypindolol bound to receptor and precipitation was abolished by the addition of unlabeled isoproterenol. Two-dimensional gel analysis and partial tryptic digestion showed the peptide precipitated by anti-Id to have identical PI and degradation fragments as the β-adrenergic receptor.[38] Thus, reovirus Type 3 is a further example of a virus which uses physiological receptors for cell attachment.[83]

IX. ANTI-ID AS AN APPROACH TO IDENTIFYING RECEPTORS ON IMMUNOCOMPETENT CELLS AND THEIR PRODUCTS

Anti-Id antibodies were also used to characterize a number of receptors important for immunocompetent cell-cell communication and for effector lymphocyte functions. Two examples are cited here. Labris and Ross[84] found that an anti-Id to Factor H ($β_1$ H globulin), an important control protein of the alternative pathway of complement activation, bound to B cell and sheep RBC coated with the active fragment of C3, but not to T lymphocytes. These findings may have been due to the fact that the anti-Id recognizes common antigenic determinants shared by antifactor H, C3b, and membranes of B cells. The anti-Id, however, precipitated from intrinsically labeled B cells, a structure distinct from that known for C3b, suggesting that the receptor for Factor H is not C3b, but shared the structure of H binding site with C3b. The Factor H anti-Id showed a wide spectrum of C3 binding activity resulting in cleavage of fluid-phase and bound C3b.

Anti-Id may further be used to physically isolate complementary receptors. Iversen et al.[85] thus used anti-Id to isolate an antigen-specific T-cell-derived factor. Sheep were immunized with antigen binding material obtained from BALB/c mice 4 days after immunization with sheep RBC. After appropriate absorption, the anti-Id reagent was used to isolate T-cell factors from BALB/c hyperimmune serum. The T-cell factors bound sheep glycophorin specifically and was found to share constant region determinants with other antigen-specific T-cell factors.

X. PERSPECTIVE

The examples of anti-Id antibodies as probes for receptor structure and function discussed here are by no means comprehensive, but were meant to illustrate the power of this approach in raising antireceptor antibodies specific to a variety of ligands, to illustrate the connectivity of the network for "internal image" Id and its possible implications for autoimmune anti-receptor diseases. It is beyond the scope of this chapter to discuss "internal image" anti-Id as potential vaccines.

We have previously discussed[14] the potential of anti-Id in characterizing and isolating receptors for which ligands are well characterized. The use of monoclonal Ab_1 and Ab_2 could yield unlimited quantities of reagents.[78] Conceptually, this approach has, however, been superceded by the auto-anti-idiotypic approach wherein auto-anti-Id are screened for their ability to bind to receptors or at least well-characterized antiligand antibodies.[62,63] The latter approach will allow the selection of antireceptor antibodies even for receptors whose structure is unknown.[63] Anti-Id may also be used to ask questions about hormone domains which interact with the receptor.[24,49] In theory, it should be possible to raise such antibodies against the battery of monoclonals specific for the hormone-specific domains. Anti-Id thus obtained should bind to the receptor epitopes complementary to specific hormone domains. Such anti-Id would be useful in investigating the conformational changes in the hormone necessary for high affinity bindng and triggering of the receptor.

As I have shown, anti-Id antibodies could be used for immunoaffinity purification of receptors, their immunoprecipitation, identification on protein blots, and immunocytochemistry and the study of surface modulation of the receptor. They clearly have a potential for a new type of receptor radioassay. Anti-Id may be potentially useful in screening expression libraries, for purification of protein for sequencing, and oligonucleotide probes selecting protein-specific mRNA by precipitation of polysomes and verifying fidelity of translation products.[85] The heavy glycosylation of membrane receptors and the role of carbohydrates in ligand recognition function may, however, limit their usefulness in this respect.

Could antireceptor antibodies arise as a result of the release of "internal image" anti-Id from immunoregulatory influences? If so, how can we explain the presence in the sera of patients with some antireceptor autoimmune diseases of antibodies directed against cell-type-specific organelles other than receptor? Studies reviewed here suggest that the answer to the first question is yes. This postulate requires that "internal image" of the ligand triggers the appearance of antireceptor antibodies as a result of network dysregulation. This "initial" antireceptor antibody may influence receptor function, as well as injure the receptor and the cell that bears it, propagating an autoaggressive response to receptor as well as appearance of antibodies to other cellular components. Another possibility is that parallel Id with antireceptor activity and activity against other cellular organelles may be activated. There is little evidence to support the notion that such natural antibodies which have been released from network control would account for the multiple varieties of antibodies because of their multispecificity. The restricted heterogeneity of several spontaneous antireceptor antibodies[11] is not inconsistent with the "initial sin" hypothesis above.

Lastly, in the context of hormone receptors, I note that Id/anti-Id complimentarity so prominent in the immune network concept has been extended to hormones and their receptors.[14] Thus, it has been found for a number of peptide hormones that the protein coded by the RNA complimentary to the mRNA that codes for the ligand, codes for relevant receptor domain.[86-88] This approach has been used to raise anticorticotropin receptor antibody, whereby its receptors have been characterized and purified.[86,88]

ACKNOWLEDGMENT

I wish to thank Dr. Ted Lo for sharing with me the results of unpublished experiments from his laboratory and Miss Gail O'Brien for expert secretarial assistance.

REFERENCES

1. **Jerne, N. K.,** Towards a network theory of the immune system, *Ann. Immunol. (Inst. Pasteur),* 125C, 373, 1974.
2. **Jerne, N. K.,** The immune system: a web of V-domains, in *Harvey Lectures,* Vol. 70, Academic Press, New York, 1976, 93.
3. **Jerne, N. K.,** The generative grammer of the immune system, *EMBO J.,* 4, 847, 1985.
4. **Vaz, N., Martinez, A. C., and Coutinho, A.,** The uniqueness and boundaries of the idiotypic self, in *Idiotypy in Biology and Medicine,* Kohler, H., Urbain, J., and Cazenave, P. A., Eds., Academic Press, Orlando, 1984, 43.
5. **Köhler, H.,** The immune network revisited, in *Idiotypy in Biology and Medicine,* Kohler, H., Urbain, J., and Cazenave, P. A., Eds., Academic Press, Orlando, 1984, 3.
6. **Bona, C. and Moran, T.,** Idiotype vaccines, *Ann. Inst. Pasteur/Immunol.,* 136, 299, 1985.
7. **Roth, C., Rocca-Serra, J., Somme, G., Fougereau, M., and Theze, J.,** Gene repertoire of the anti-poly (Glu Ala Tyr) (GAT) immune response: comparison of V, V, and D regions used by anti-GAT antibodies and monoclonal antibodies produced after anti-idiotypic immunization, *Proc. Natl. Acad. Sci. U.S.A.,* 82, 4788, 1985.
8. **Bona, C. A., Victor-Kobrin, C., Manheimer, A. J., Bellon, B., and Rubenstein, L. J.,** Regulatory arms of the immune networks, *Immunol. Rev.,* 79, 25, 1984.
9. **Erlanger, B. F.,** Anti-idiotypic antibodies: what do they recognize?, *Immunol. Today,* 6, 10, 1985.
10. **Roitt, I. M., Thanavala, Y. M., Male, D. K., and Hay, F. C.,** Anti-idiotypes as surrogate antigens: structural considerations, *Immunol. Today,* 6, 265, 1985.
11. **Farid, N. R., Briones-Urbina, R., and Bear, J. C.,** Graves' disease — the thyroid stimulating antibody and immunological networks, *Mol. Aspects Med.,* 6, 355. 1983.
12. **McNamara, M. and Kohler, H.,** Regulatory idiotopes. Induction of idiotype-recognizing helper T cells by free light and heavy chains, *J. Exp. Med.,* 159, 623, 1984.
13. **Janeway, C. A., Jr.,** Idiotypes to their expression, *Immunol. Today,* 2, i, 1981.
14. **Farid, N. R. and Lo, T. C. Y.,** Antiidiotypic antibodies as probes for receptor structure and function, *Endocrinol. Rev.,* 6, 1, 1985.
15. **Mostov, K. E., Friedlander, M., and Blobel, G.,** The receptor for transepithelial transport of IgA and IgN contains multiple immunoglobulin-like domains, *Nature (London),* 308, 37, 1984.
16. **Williams, A. F.,** The immunoglobulin family takes shape, *Nature (London),* 308, 812, 1984.
17. **Strosberg, A. D., Couraud, P. O., and Schreiber, A.,** Immunological studies of hormone receptor: a two-way approach, *Immunol. Today,* 2, 75, 1981.
18. **Bona, B. A., Findley, S., Walters, S., and Kunkel, H. G.,** Anti-immunoglobulin antibodies. III. Properties of sequential anti-idiotypic antibodies to heterologous anti-α-globulins. Detection of reactivity of anti-idiotypic antibodies with epitopes of Fc fragments (homobodies) and with epitopes and idiotopes (epibodies), *J. Exp. Med.,* 156, 96, 1982.
19. **Sege, K. and Petersen, P. A.,** Use of anti-idiotypic antibodies as cell-surface receptor probes, *Proc. Natl. Acad. Sci. U.S.A.,* 75, 2443, 1978.
20. **Sege, K. and Petersen, P. A.,** Anti-idiotypic antibodies against anti-vitamin A transporting react with pre-albumen, *Nature (London),* 271, 167, 1978.
21. **Khokher, M. A., Danoda, P., Janah, S., and Coulston, G.,** Insulin-like stimulatory effect of human immunoglobulin G on adipocyte lipogenesis, *Diabetes,* 30, 1068, 1981.
22. **Gaulton, G. N. and Greene, M. I.,** Idiotypic mimicry of biological receptor, *Annu. Rev. Immunol.,* 4, 253, 1984.
23. **Elias, D., Maron, R., Cohen, I. R., and Shecter, Y.,** Mouse antibodies to the insulin receptor developing spontaneously as anti-idiotypes. II. Effects on glucose homeostasis and insulin receptor, *J. Biol. Chem.,* 259, 6416, 1984.
24. **Schechter, Y., Elias, D., Maron, R., and Cohen, I. R.,** Mouse antibodies to insulin receptor developing spontaneously as anti-idiotypes. I. Characterization of the antibodies, *J. Biol. Chem.,* 259, 6411, 1984.

25. **Maron, R., Elias, D., de Jongh, B. M., Bruining, G. J., van Rood, J. J., Shechter, Y., and Cohen, I. R.,** Autoantibodies to the insulin receptor in juvenile onset insulin-dependent diabetes, *Nature (London),* 303, 817, 1983.

26. **Palmer, J. P., Asplin, C. M., Clemons, P., Lyen, K., Tatpati, O., Raghu, P. K., and Paquette, T. L.,** Insulin antibodies in insulin-dependent diabetics before insulin treatment, *Science,* 222, 1337, 1983.

27. **Farid, N. R.,** Immunologic aspects of diabetes, in *Nutrition and Immunology,* Chandra, R. K., Ed., Alan R. Liss, New York, in press.

28. **Schreiber, A. B., Couraud, P. O., Andre, C., Vray, B., and Strosberg, A. D.,** Anti-alprenolol anti-idiotypic antibodies bind to β-adrenergic receptors and modulate catecholamine-sensitive adenylate cyclase, *Proc. Natl. Acad. Sci. U.S.A.,* 77, 7385, 1980.

29. **Homcy, C. J., Rockson, S. G., and Haber, G.,** An antiidiotypic antibody that recognizes the β-adrenergic receptor, *J. Clin. Invest.,* 69, 1147, 1982.

30. **Rockson, S. G., Homcy, C. J., and Haber, E.,** Antialprenolol antibodies in the rabbit. A new probe for the study of β-adrenergic receptor interactions, *Circ. Res.,* 46, 808, 1980.

31. **Couraud, P. O., Schreiber, A. B., Delavier-Klutchko, C., Andre, C., Durieu-Trautmann, O., Vray, B., Schmutz, A., and Strosberg, A.,** Immunological studies of adrenergic catecholamine receptors: a two way approach, in *Protides of Biological Fluids,* Vol. 29, Peeters, H., Ed., Pergamon Press, Oxford, 1981, 493.

32. **Strosberg, A. D., Couraud, P. O., Durieu-Trautmann, O., and Delavier-Klutcko, C.,** Biochemical and immunochemical analysis of β-adrenergic receptor adenylate cyclase complexes, *Trends Pharmacol. Sci.,* 3, 282, 1982.

33. **Strosberg, A. D.,** Anti-idiotype and anti-hormone receptor antibodies, *Springer Semin. Immunopathol.,* 6, 67, 1983.

34. **Couraud, P.-O., Lu, B. U., and Strosberg, A. D.,** Cyclic anti-idiotypic response to anti-hormone antibodies due to neutralization of autologous anti-anti-idiotype antibodies that bind hormones, *J. Exp. Med.,* 157, 1369, 1983.

35. **Kelsoe, G. and Cerny, J.,** Reciprocal expansions of idiotypic and anti-idiotypic clones following antigenic stimulation, *Nature (London),* 279, 333, 1979.

36. **Guillet, J. G., Kaveri, S. V., Durie, U. O., Delavier, C., Hoebeke, J., and Strosberg, A. D.,** β-Adrenergic agonist activity of a monoclonal antiidiotypic antibody, *Proc. Natl. Acad. Sci. U.S.A.,* 82, 1781, 1985.

37. **Co, M. S., Gaulton, G. N., Fields, B. N., and Greene, M. I.,** Isolation and characterization of the mammalian reovirus type 3 cell surface, *Proc. Natl. Acad. Sci. U.S.A.,* 82, 1494, 1985.

38. **Co, M. S., Gaulton, G. N., Tominga, A., Homcy, C. J., Fields, B. N., and Green, M. I.,** Structural similarities between mammalian β-adrenergic and reovirus type 3 receptors, *Proc. Natl. Acad. Sci. U.S.A.,* 82, 5315, 1985.

39. **McKenzie, J. M. and Zakarija, M.,** LATS in Graves' disease, *Recent Prog. Horm. Res.,* 33, 29, 1978.

40. **Islam, M. N., Pepper, B. M., Briones-Urbina, R., and Farid, N. R.,** Biological activity of anti-thyrotropin anti-idiotypic antibody, *Eur. J. Immunol.,* 13, 57, 1983.

41. **Knopp, J., Stolc, V., and Tong, W.,** Evidence for the induction of iodide transport in bovine thyroid cells treated with thyroid-stimulating hormone or dibutyryl cyclic adenosine 3',5'-monophosphate, *J. Biol. Chem.,* 245, 4403, 1970.

42. **Williams, J. A. and Malayan, S. A.,** Effects of TSH on iodide transport by mouse thyroid lobes in vitro, *Endocrinology,* 97, 162, 1975.

43. **Biro, J.,** Specific binding of thyroid stimulating hormone by human globulin, *J. Endocrinol.,* 88, 339, 1981.

44. **Biro, J.,** Thyroid-stimulating antibodies in Graves' disease and the effect of thyrotropin-binding globulins on their determinations, *J. Endocrinol.,* 92, 175, 1982.

45. **Holmberg, D. and Coutinho, A.,** Natural antibodies and autoimmunity, *Immunol. Today,* , 6, 356, 1985.

46. **Islam, M. N. and Farid, N. R.,** Structure of the porcine thyrotropin receptor: a 200 kilodalton hetero-complex, *Experimentia,* 41, 18, 1984.

46a. **Farid, N. R., Briones-Urbina, R., and Islam, M. N.,** Anti-idiotypic antibodies are probes for hormone-receptive interaction, *Can. J. Biochem. Cell Biol.,* 62, 1255, 1984.

47. **Islam, M. N., Briones-Urbina, R., Bako, G., and Farid, N. R.,** Both TSH and thyroid stimulating antibody of Graves' disease bind to an Mr 197,000 holoreceptor, *Endocrinology,* 113, 436, 1983.

48. **Bako, G., Islam, M. N., and Farid, N. R.,** Photoaffinity labelling of the porcine thyrotropin receptor, *Clin. Invest. Med.,* 8, 152, 1984.

49. **Briones-Urbina, R., Islam, M. N., Ivanyi, J., and Farid, N. R.,** Use of anti-idiotypic antibodies as probes for the interaction of TSH subunits with its receptor, *J. Cell Biochem.,* 34, 151, 1987.

50. **Pierce, J. G. and Parsons, T. F.,** Glycoprotein hormones: structure and function, *Annu. Rev. Biochem.,* 50, 465, 1981.

51. **Holmes, N. J. and Parham, P.,** Enhancement of antibodies against HLA-A2 due to antibody bivalency, *J. Biol. Chem.,* 258, 1580, 1983.
52. **Milius, R. P., Midgley, A. R., Jr., and Birkens, S.,** Preferential masking by the receptor of immunoreactive sites on the subunit of human choriogonadotropin, *Proc. Natl. Acad. Sci. U.S.A.,* 80, 7375, 1983.
53. **Valente, W. A., Vitti, P., Yavin, Z., Yavin, E., Rotella, C. N., Grollman, E. F., Toccafondi, R. S., and Kohn, L. D.,** Monoclonal antibodies to the thyrotropin receptor: stimulating and blocking antibodies derived from the lymphocytes of patients with Graves' disease, *Proc. Natl. Acad. Sci. U.S.A.,* 79, 6680, 1982.
54. **Islam, M. N., Tuppal, R., Hawe, B. S., Briones-Urbina, R., and Farid, N. R.,** The thyroid "microsomal" antigen is an epitope on the thyrotropin receptor, *J. Cell. Biochem.,* 31, 107, 1986.
55. **Briones-Urbina, R. and Farid, N. R.,** Control of the biosynthesis and turnover of the thyrotropin receptor, manuscript submitted.
56. **Baker, J. R., Lukes, Y. G., and Burman, K. D.,** Production, isolation and characterization of rabbit anti-idiotypic antibodies directed against human anti-thyrotropin receptor antibodies, *J. Clin. Invest.,* 74, 488, 1984.
57. **Beall, G. N., Rapoport, B., Chopra, I. J., Solomon, D. K., and Kruger, S.,** Rabbits immunized with thyroid stimulating hormone produce thyroid stimulating antibodies, *J. Clin. Invest.,* 75, 1435, 1985.
58. **Moyle, W. R., Ehrlich, P. H., and Canfield, R. E.,** Use of monoclonal antibodies to subunits of human chorionic gonadotropin to examine the orientation of the hormone in its complex with receptor, *Proc. Natl. Acad. Sci. U.S.A.,* 79, 2245, 1982.
59. **Amit, T., Barkey, R. J., Gavish, M., and Youdim, M. B. H.,** Anti-idiotypic antibodies raised against anti-prolactin (PRL) antibodies recognize the PRL receptor, *Endocrinology,* 118, 835, 1986.
60. **Amit, T., Gavish, M., Barkey, R. J., and Youdim, M. B. H.,** Anti-idiotypes to PRL antibody have PRL-like activity and bind to the PRL receptor, Program of the 7th Int. Congr. of Endocrinology, Abstr. 202, Quebec City, PQ, July 1 to 7, 1984.
61. **Wassermann, N. H., Penn, A. S., Freimuth, P. I., Treptow, N., Wentzel, S., Cleveland, W. L., and Erlanger, B. F.,** Anti-idiotypic route to anti-acetylcholine receptor antibodies and experimental myasthenia gravis, *Proc. Natl. Acad. Sci. U.S.A.,* 79, 4810, 1982.
62. **Cleveland, W. L., Wassermann, N. H., Sarangarajan, R., Penn, A. S., and Erlanger, B. F.,** Monoclonal antibodies to the acetycholine receptor by a normally functioning auto-anti-idiotypic mechanism, *Nature (London),* 305, 56, 1983.
63. **Cleveland, W. L., Wassermann, N. H., Penn, A. S., Ku, H. H., Hill, B. L., Sarangarajan, R., and Erlinger, B. F.,** Idiotypic routes to monoclonal anti-receptor antibodies, in *Monoclonal Antibodies and Cancer Therapy,* UCLA Symp. on Molecular and Cellular Biology, New Series, Vol. 27, Reisfeld, R. A., and Sell, S., Eds., Alan R. Liss, New York, 1985, 345.
64. **Ng, D. S. and Isom, G. E.,** Binding of anti-morphine anti-idiotypic antibodies to opiate receptors, *Eur. J. Pharmacol.,* 102, 187, 1984.
65. **Glasel, J. A. and Myers, W. E.,** Rabbits anti-idiotypic antibodies raised against monoclonal anti-morphine IgG block and opiate receptor sites, *Life Sci.,* 36, 2523, 1985.
66. **Gramsch, C., Meo, T., Riethmuller, G., and Herz, A.,** *J. Neurochem.,* 40, 1220, 1983.
67. **Meo, T., Gramsch, C., Inan, R., Hollt, V., Weber, E., Herz, A., and Reithmuller, G.,** *Proc. Natl. Acad. Sci. U.S.A.,* 80, 4084, 1983.
67a. **Schulz, R. and Gramsch, C.,** Polyclonal anti-idiotypic opioid receptor antibodies generated by the monoclonal γ-endorphin antibody 3-E7, *Biochem. Biophys. Res. Commun.,* 132, 658, 1985.
68. **Carter-su, C., Pessin, J. E., Mora, R., Gitomer, W., and Czech, M. P.,** Photoaffinity labelling of the human erythrocyte D-glucose transporter, *J. Biol. Chem.,* 257, 5419, 1982.
69. **Duronio, V. and Lo, T. C. Y.,** Identification of hexose transport components through the use of anti-idiotypic antibodies, manuscript submitted.
70. **Koepsell, H., Menhur, H., Ducis, I., and Wissmuller, T. F.,** Partial purification and reconstitution of the Na$^+$-D-glucose cotransport protein from pig renal proximal tubules, *J. Biol. Chem.,* 258, 1888, 1983.
71. **Kay, M. M. B.,** Glucose transport protein is structurally and immunologically related to band 3 and senescent cell antigen, *Proc. Natl. Acad. Sci. U.S.A.,* 82, 1731, 1985.
72. **Mullins, R. E. and Langdon, R. G.,** Maltosyl isothiocyanate: an affinity label for the glucose transporter of the human erythrocyte membrane. II. Identification of the transporter, *Biochemistry,* 11, 1199, 1980.
73. **Lo, T. C. Y.,** The molecular mechanisms of substrate transport in gram-negative bacteria, *Can. J. Biochem.,* 57, 289, 1979.
74. **Lo, T. C. Y. and Bewick, M. A.,** Use of a non-penetrating substrate analogue to study the molecular mechanism of the outer membrane dicarboxylate transport system in, *E. coli* K12, *J. Biol. Chem.,* 256, 5511, 1981.
75. **Lo, T. C. Y. and Bewick, M. A.,** The molecular mechanisms of carboxylic acid transport in *E. coli:* the role and orientation of the two membrane bound dicarboxylate binding proteins, *J. Biol. Chem.,* 254, 7826, 1978.

76. **Marasco, W. A. and Becker, E. L.,** Anti-idiotype as antibody against the formyl peptide chemotaxis receptor of the neutrophil, *J. Immunol.,* 128, 963, 1982.
77. **Marasco, W. A., Charles, J., and Ward, P. A.,** A search for "internal image" anti-idiotypic antibodies: a subpopulation of anti-f Met-Leu-Phe anti-idiotypic antibodies binds to the formyl peptide chemotaxis receptor of the neutrophil and to several genetically unrelated anti-f Met-Leu-Phe antibodies, in *Monoclonal Antibodies and Cancer Therapy,* UCLA Symp. on Molecular and Cellular Biology, New Series, Vol. 27, Reisfeld, R. A. and Sell, S., Eds., Alan R. Liss, New York, 1985, 327.
78. **Noseworthy, J. H. and Greene, M. I.,** Studies on idiotypes shared by neuronal and lymphoid cells, in *Idiotypy in Biology and Medicine,* Kohler, H., Urbain, J., and Cazenave, P. A., Ed., Academic Press, Orlando, 1984, 303.
79. **Ertl, H. C. J., Greene, M. I., Noseworthy, J. H., Fields, B. N., Nepom, G. T., Spriggs, D. R., and Fineberg, R. W.,** Identification of idiotypic receptors on reovirus-specific cytolytic T cells, *Proc. Natl. Acad. Sci. U.S.A.,* 79, 7479, 1982.
80. **Noseworthy, J. H., Fields, B. N., Dichter, M. K., Sobotka, C., Pizer, E., Perry, L. A., Nepom, J. T., and Greene, M. I.,** Cell receptors for the mammalian reovirus. I. Syngeneic monoclonal anti-idiotypic antibody identifies a cell surface receptor for reovirus, *J. Immunol.,* 131, 2533, 1983.
81. **Kauffman, R. S., Noseworthy, J. H., Nepom, J. T., Fineberg, R., Fields, B. N., and Greene, M. I.,** Cell receptors for the mammalian reovirus. II. Monoclonal anti-idiotypic antibody blocks viral binding to cells, *J. Immunol.,* 131, 2539, 1983.
82. **Sharpe, A. H., Gaulton, G. N., McDade, K. K., Fields, B. N., and Greene, M. I.,** Syngeneic monoclonal antibodies can induce cellular immunity to reovirus, *J. Exp. Med.,* 160, 1195, 1984.
83. **Sharpe, A. H. and Fields, B. N.,** Pathogenesis of viral infections. Basic concepts derived from the reovirus model, *N. Engl. J. Med.,* 312, 486, 1985.
84. **Labris, J. D. and Ross, G. C.,** Characterization of the lymphocyte membrane receptor for factor H (β_1, H-globulin) with an antibody to anti-factor H-idiotype, *J. Exp. Med.,* 155, 1400, 1982.
85. **Gaulton, G. N., Co, M. S., Royer, H. D., and Greene, M. I.,** Anti-idiotypic antibodies as probes of cell surface receptors, *Mol. Cell. Biochem.,* 65, 5, 1984.
86. **Bost, K. L., Smith, E. M., and Blalock, J. E.,** Similarity between the corticotropin (ACTH) receptor and a peptide encoded by an RNA that is complementary to ACTH mRNA, *Proc. Natl. Acad. Sci. U.S.A.,* 82, 1372, 1985.
87. **Bost, K. L., Smith, E. M., and Blalock, J. E.,** Regions of complementarity between the messenger RNA's for epidermal growth factor, transferrin, interleukin-2 and their receptor, *Biochem. Biophys. Res. Commun.,* 128, 1373, 1985.
88. **Bost, K. L. and Blalock, J. E.,** Molecular characterization of a corticotropin (ACTH) receptor, *Mol. Cell. Endocrinol.,* 44, 1, 1986.
89. **Farid, N. R., Fahraeus-Van Ree, Hawe, B., Thompson, C., and Kozma, L.,** unpublished.
90. **Farid, N. R., Davies, T. F., and Hawe, B. S.,** unpublished.
91. **Moyle, W. R.,** Department of Obstet/Gyn, UMDNJ-Rutgers Medical School, personal communications.
92. **Walker, C. and Lo, T. C. Y.,** unpublished data.

Chapter 7

CROSS-REACTING AND REGULATORY IDIOTYPES ON AUTOANTIBODIES: EFFECT ON REGULATION OF AUTOIMMUNITY*

M. Zanetti**, D. Glotz***, and J. Rogers

TABLE OF CONTENTS

*   This is publication number 94 from Medical Biology Institute, 11077 North Torrey Pines Road, La Jolla, Calif. 92037. This work was supported by NSF grant DCB-8502601.
** Dr. Maurizio Zanetti is a Scholar of the Leukemia Society of America, Inc.,
***Dr. Denis Glotz is supported by grants from the Foundation pour la Recherche Medicale, Ministere des Relations Exterieures and Ministere de l'Industrie et de la Recherche, of France.

I. INTRODUCTION

During the past decade ample evidence has accumulated to suggest that idiotypic determinants on antibody/cell variable (V) regions are sites for recognition and regulation of immune responsiveness[1] not only to exogenous or alloantigens, but also to self-antigens. In both instances, perturbation of the immune system by idiotype (Id) or anti-idiotype (anti-Id) is usually followed by a modification of the immune response either at the humoral or cellular level, or both levels. Since the immune network can be set in either of two functional states, suppressive or enhancing, an imbalance in the delicate dynamic equilibrium between Id and anti-Id may affect self-tolerance and autoreactivity. Thus, it has been shown that immunologic self-tolerance can be circumvented by anti-Id antibodies alone[2,3] and, conversely, the appropriate administration of anti-Id antibodies can specifically down regulate an autoantibody response.[4-6]

That the immune system is physiologically acting on a self-recognition mode[7] and that Id are part of this process[1] is now accepted. Additionally, since Id can be the immunologic representation of nominal antigens, completeness is a distinctive feature of the immune system, i.e., all the possible antigens are or can be represented within the universe of V regions.[8] Furthermore, it is proposed that the immune system is essentially self-referential[9] and, whenever self-antigens are involved, self-centered.[10] With this in mind, we have introduced the concept of an *autoimmune network* as the comprehensive framework for antiself-reactivity based on the positive and negative influences of receptor/antireceptor, self-specific interactions within the Id network à la Jerne.[11]

The dynamics of the autoimmune network are subjected to regulation by nominal autoantigens, Id, and anti-Id. While the role of self-antigens in the induction of autoreactivity or establishment of tolerance has been extensively studied (for review, see Reference 12), little is known about the functional role exerted by Id and their complementary anti-Id in each instance. Evidence exists to suggest that Id may not all be equal, and quantitative/qualitative differences among the multitude of Id one possesses or produces in response to a given antigen do exist. Id shared by the majority of antibodies to one antigen — crossreacting Id (CRI) — usually are dominant, in a quantitative sense, within that immune response. On the other hand, Id which primarily regulate — regulatory Id (ReI) — are not necessarily dominant in the serum of immunized organisms. Although this concept is still in its infancy, it is proposed that the main characteristics of ReI are to be autoimmunogenic, activate B and T cells, and determine the magnitude and possibly the nature of the immune response.[13] Studies pertaining to functions of ReI will be presented herein and will focus on our recent work on a ReI denoted Id62, which is borne on autoantibodies to a classical autoantigen (Tg).

CRI on V regions for self-antigens have now been demonstrated in numerous systems, in animals and in humans, both at the monoclonal and polyclonal level (for review, see Reference 14). ReI have so far only been demonstrated in experimental systems[11,15] and their counterpart in humans has yet to be demonstrated. Both CRI and ReI, which we believe are highly conserved structures of the autoimmune repertoire,[11] are the best candidates for regulation of the autoimmune network. However, whether the mechanism through which they are involved in immune regulation is the same is not clear as yet. Besides this theoretical view, the notion that the autoimmune network may be spontaneously operational also stems from the direct evidence for auto-anti-Id in animals and in humans (reviewed in Reference 16).

It is the object of this communication to discuss separately regulation of the autoimmune network via CRI and ReI. We will refer to those experiments performed in our laboratory over the last few years which constitute an attempt to understand how the autoimmune network works and how strategies of therapeutic intervention for autoimmune diseases can be eventually designed.

Table 1
**CORRELATION BETWEEN SUPPRESSION OF TUBULO-
INTERSTITIAL NEPHRITIS AND LEVELS OF
AUTOANTIBODIES REACTING WITH GLYCOPROTEINS OF
AUTOLOGOUS TBM ANTIGEN IN RATS TREATED WITH
SOLUBLE ANTI-CRI ANTIBODY**

Treatment	No. rats	Autoantibodies (μg/mℓ)[a]	Cortical involvement (%)[b]			
			0	<25	25—50	50—100
Saline	6	>3	0	3	0	3
Control antibody	7	>3	0	2	2	3
Anti-CRI antibody	5	>3	1	1	1	2
	5	<3	3	2	0	0

[a] Autoantibodies quantitated by solid-phase radioimmunoassay.
[b] Cortical involvement by lymphomonocytic infiltration assessed by light microscopy on whole kidney sections obtained along the longitudinal axis.

Modified from Zanetti, M., Mampaso, F., and Wilson, C. B., *J. Immunol.*, 131, 1268, 1983.

II. REGULATION OF AUTOIMMUNITY BY ANTIBODIES TO CRI ON AUTOANTIBODIES

Attempts to regulate the production of autoantibodies expressing CRI have been made in several animal models of autoimmunity. Experimentally induced or spontaneously occurring autoimmune responses have both been used. One model of organ-specific autoimmune disease utilized in several laboratories, including our own, is the experimental tubular interstitial nephritis in rodents. In this disease, autoantibodies to antigens of the kidney tubular basement membrane (TBM) are considered to be of causative importance. In the Brown-Norway rats, the antibody response to TBM is complex and serum antibodies react with multiple moieties of the TBM, collagenous and noncollagenous.[17] In contrast to the serum, the majority of autoantibodies recovered by elution from diseased kidneys react with the glycoprotein fraction of autologous TBM. Using a rabbit anti-Id prepared against autoantibodies eluted from diseased kidneys, we showed that TBM-reactive autoantibodies and B lymphocytes from immunized rats bear a CRI.[6] A single i.p. injection of the heterologous anti-CRI antibody prior to induction of the disease markedly suppressed the production of autoantibodies to TBM.[6] Interestingly, the suppression was confined to autoantibodies to TBM glycoprotein moieties, i.e., the same specificity of nephritogenic antibodies. As a consequence, the lymphomonocytic infiltration of the cortical interstitium typical of this nephritis decreased in parallel with the degree of autoantibody suppression (Table 1).

From a practical standpoint, however, it is of greater importance to assess the effect of anti-CRI during ongoing, spontaneous autoimmunity. One such experiment was done utilizing the spontaneous autoimmune thyroiditis of Buffalo (BUF) rats. These rats develop an autoimmune thyroiditis very similar to Hashimoto's thyroiditis in humans associated with circulating autoantibodies to Tg. Spontaneous autoantibodies to Tg in the BUF strain possess a CRI which is found in 70% of randomly chosen animals and constitutes 11 to 47% of the autoantibodies.[5] The same CRI can also be demonstrated in the sera of BUF rats immunized with homologous Tg in CFA.[18] At the cellular level, CRI-positive lymphocytes can be detected in the spleen of BUF rats with either spontaneous or experimentally induced thyroiditis.[18] Repeated weekly i.p. injections of anti-CRI antibody into rats with high levels of circulating autoantibodies to Tg significantly ($p < 0.05$ to 0.01) decreased circulating auto-

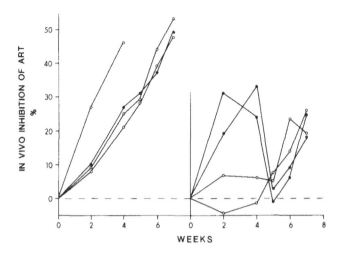

FIGURE 1.   In vivo inhibition of production of autoantibodies to rat thyroglobulin (ART) in individual BUF rats with spontaneous, ongoing autoimmune thyroiditis, following repeated i.p. injections of a rabbit anti-CRI antibody. Each point represents the percent change of the binding of ^{125}I-labeled Tg by sera of treated rats from the pretreatment value. Open circles (O — O) refer to 6-month-old rats with neonatal thymectomy, and closed circles (● — ●) refer to retired breeders ($\geqq$1-year-old) rats. Rats of both groups had comparable high levels of circulating autoantibodies at the beginning of the experiment. (Modified from Zanetti, M. and Bigazzi, P. E., *Eur. J. Immunol.*, 11, 187, 1981. With permission.)

antibodies from the pretreatment value in each case and throughout the duration of the experiment (7 weeks). The maximum in vivo suppression varied between 18 and 53%. However, when individually analyzed, rats could be divided into two homogeneous groups: one in which the inhibition ranged from 46 to 53%; the other in which lower (18 to 26%) inhibition occurred (Figure 1). Histologic examination failed to show any meaningful improvement of the histologic lesions of thyroiditis. Anti-CRI-treated rats that underwent suppression no longer displayed the circulating CRI on autoantibodies to Tg.

Although in both experiments only partial suppression was documented, the evidence is that the predominant effect of treatment by soluble anti-CRI antibodies is the down regulation of autoreactive clones. Whether suppression by anti-CRI occurred because of direct functional inactivation of autoreactive B lymphocytes or through the elicitation of T suppressor cells remains to be determined.

III. REGULATION OF AUTOIMMUNITY BY ANTIBODIES TO ReI

Studies performed in our laboratory have focused on regulation of the autoimmune response to Tg using Id62 as putative ReI. Initial experiments were designed to explore the effect of immunization with antibodies to Id62. To this end, adult BALB/c mice were immunized with rabbit anti-Id62 antibody alone and subsequently tested for Id expression and production of antibodies to Tg. Figure 2 shows that more than 50% of mice develop autoantibodies to Tg as detected ELISA on murine Tg-coated polyvinyl microtiter wells. At the same time, marked levels of Id62-positive antibodies (Id') were detected in all but one mouse injected with anti-Id. The relationship between Id' molecules and autoantibodies to Tg was elucidated using a panel of Tg-reactive monoclonal antibodies derived from spleen lymphocytes of a mouse immunized with the anti-Id62 antibody. When such hybridoma antibodies were studied for Id expression, they were found to be very similar to prototype antibody 62 as they could inhibit the Id-anti-Id ELISA binding with similar stoichiometry.

As demonstrated, immunization with anti-Id alone can, at least in certain instances,

BALB/c MICE IMMUNIZED WITH PURIFIED RABBIT ANTI-Id62 (ANTI-ANTI-THYROGLOBULIN) ANTIBODIES PRODUCE ANTIBODIES REACTIVE WITH HOMOLOGOUS THYROGLOBULIN AND Id62 (Id')

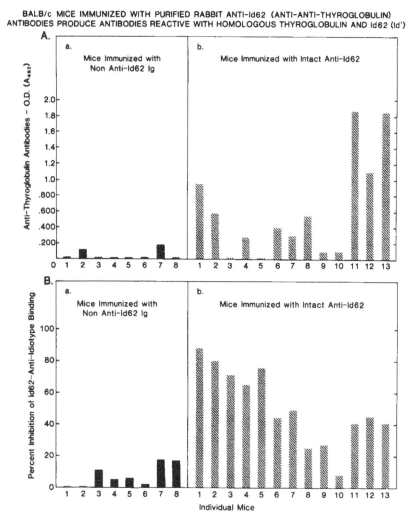

FIGURE 2.   Detection of autoantibodies to Tg and Id' in mice immunized with purified anti-Id (anti-[anti-Tg] antibodies). Eight-week-old BALB/c mice were first tolerized to rabbit gamma globulins. One week after tolerization, mice were immunized i.p. with 25 μg of affinity-purified rabbit anti-Id62 antibodies (right panels) in CFA. Anti-Id62 antibodies were purified by affinity chromatography from a Sepharose®-4B immunoadsorbent coated with an Id62-positive Tg-specific monoclonal antibody, mAb 1.15, which had been purified on a protein A/Sepharose® 4B column. This strategy was adopted to minimize the copurification of antibodies to minor idiotypic determinants on mAb 62. Ig of the pass-through fraction, i.e., lacking anti-idiotypic activity, were used to immunize control mice (left panels). Mice were boosted after 2 weeks with an i.p. injection of 5 μg of the same antibody in alum. Sera were individually collected 10 days after the booster injection and tested. Antibodies to Tg (upper panels) were detected by ELISA on microtiter plates coated with mouse Tg using a 1:500 serum dilution. The results are expressed as absorbance at 492 nm. Id' was detected (bottom panels) by competitive inhibition of the ELISA binding of Id62 to microtiter wells coated with homologous rabbit anti-Id62 antibody by individual mouse sera (1:50 final dilution). The results are expressed as percent inhibition. (From Zanetti, M., Rogers, J., and Katz, D. H., *J. Immunol.*, 133, 240, 1984. With permission.)

specifically circumvent self-tolerance and lead to production of autoantibodies. Autoantibodies thus produced were found to bear the same Id (Id62) expressed on the antibody originally used to generate the anti-Id antibody. A possibility would be that the particular anti-Id62 antibody used functioned as the internal image of a Tg-related epitope. In favor of this hypothesis is the fact that autoantibodies could be induced using the same approach,

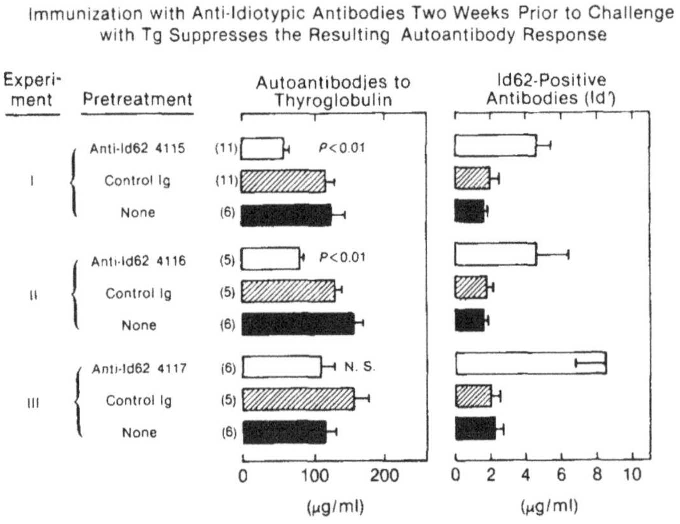

Immunization with Anti-Idiotypic Antibodies Two Weeks Prior to Challenge
with Tg Suppresses the Resulting Autoantibody Response

FIGURE 3. Immunization with anti-Id62 antibodies (50 μg/mouse in alum) suppresses the autoantibody and enhances the Id' response of mice challenged with Tg (10 μg/mouse) 2 weeks after the last booster injection with anti-Id. The numbers in parentheses refer to the number of mice in each group. Values correspond to the mean ± SEM (autoantibody to Tg) and mean ± SD (Id' antibodies) of individually tested sera collected 10 days after the second (2 weeks) booster with Tg. (From Zanetti, M., Rogers, J., and Katz, D. H., *Autoimmunity*, in press.)

not only in BALB/c mice, but also in BUF and Fischer rats,[10] hence, suggesting that this antibody, operationally speaking, mimics Tg. However, an internal image effect could not be firmly established, as autoantibodies to Tg (or Id' molecules) could not be elicited using the same antibody in guinea pigs of several strains.[10] Rather, it is possible that Id62 is a conserved ReI only in certain species, including mice and rats, but not in others such as guinea pigs. Accordingly, immunization with anti-Id62 antibody could trigger Id-positive clones (irrespective of their antigen specificity) as well as both Id62-positive/non-Tg binding clones and Id62-positive/Tg-reactive clones only in those species which possess it. Within this view, it would appear that anti-Id62 antibody is bifunctional as it can trigger clones either at their paratope (antigen binding) or at their Id. It is our experience, however, that in mice the elicitation of autoantibodies by anti-Id immunization is limited in its magnitude and duration.[10] Whether or not this indicates that beside Id-anti-Id interactions other regulatory mechanisms might be involved in the fate of anti-self-reactivity, remains to be determined.

Based on the observation that perturbation of the autoimmune network by anti-ReI can elude natural self-tolerance, preimmunization with anti-ReI antibodies should also modify the outcome of the autoimmune response induced by nominal antigen. A verification of this assumption comes from studies on anti-Id62-immunized mice subsequently challenged with Tg. A panel of rabbit-anti-Id62 antibodies was utilized which, by an ensemble of in vitro criteria, was characterized as apparently directed against a similar immunodominant epitope on prototype antibody 62.[19] As shown in Figure 3, in two out of three instances the antibody response of anti-Id-treated mice was significantly suppressed as compared to controls. At the same time, Id' molecules were found in greater excess in the serum of suppressed mice than controls. To understand this apparent paradox, the antigen-binding specificity of Id' molecules was assessed. Thus, antibodies were first affinity purified on an anti-Id62 immunoadsorbent, and then their Tg reactivity assessed quantitatively by radioimmunoassay.

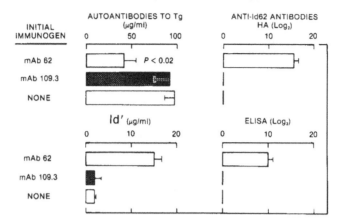

FIGURE 4. Active auto-anti-Id62 immunity induced shortly before challenge with Tg predisposes to suppression of the resulting autoantibody response. Mice were immunized with mAb 62/KLH (100 μg in alum) on day 1 and 15, then challenged with Tg (10 μg in alum) 2 weeks after the booster immunization. Assays were performed on sera collected at sacrifice, 25 days after the initial challenge with Tg. Results are expressed as mean ± SEM (autoantibodies to Tg) or mean ± SD (the remaining tests) of individual sera (five to six mice/group). (From Zanetti, M., Glotz, D., and Rogers, J., *J. Immunol.*, 137, 3140, 1986. With permission.)

It could be established that Id' molecules in suppressed mice do not, by and large, bind Tg (ratio of Id' to Tg-binding, 6:1). Therefore these experiments clearly demonstrate that during perturbation by anti-ReI, two distinct events occur. One is the expansion of the Id' compartment (parallel set), the other, suppression of autoantibodies production following challenge with nominal antigen. We confirmed these results in a subsequent experiment using a syngeneic monoclonal anti-Id62 antibody.

IV. REGULATION BY ReI

An obvious extension of the studies just described was to investigate the role of autologous immunity against the ReI itself.[20] To this end, BALB/c mice were actively immunized with Id62. Once a detectable autologous anti-Id response was established, mice were challenged with Tg. Autologous anti-Id, Id' molecules, and autoantibodies to Tg were all measured, so that a global assessment of the various compartments of the autoimmune network could be made. Figure 4 shows that fairly large amounts of auto-anti-Id antibodies were produced and, more importantly, both suppression of the autoantibody response and expansion of Id62-positive clones occurred as in previous experiments. Any possible interference by the injected Id on these results could be ruled out, since identical findings were obtained in adoptive transfers using Id62-primed spleen lymphocytes. Therefore, immunity to ReI, whichever the mode of induction may be, reproducibly caused activation of the suppressive limb of the autoimmune network.

V. ROLE OF T AND B CELLS IN ReI-DEPENDENT SUPPRESSION OF AUTOIMMUNITY

Adoptive transfers using separate B- and T-cell populations from Id62- primed mice were performed to understand the relationship between activation of the auto-anti-Id compartment, expansion of Id-positive clones, and suppression of the autoantibody response. Recipients

PRODUCTION OF AUTO-ANTI-Id62 ANTIBODIES AND SUPPRESSION OF THE
AUTOANTIBODY RESPONSE TO Tg USING SEPARATE T AND B
LYMPHOCYTES FROM DONORS WITH ACTIVE AUTO-ANTI-IDIOTYPIC IMMUNITY

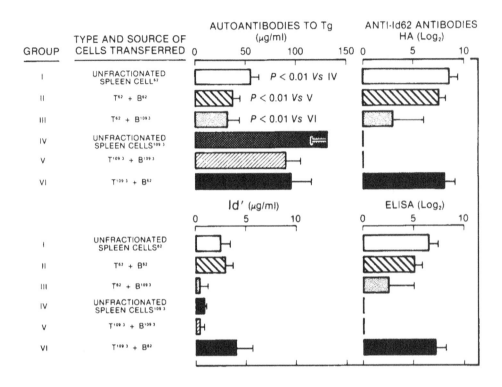

FIGURE 5.   Production of auto-anti-Id62 antibodies and suppression of the autoantibody response to Tg in mice adoptively transferred with separate T and B lymphocytes from donors with active auto-anti-Id62 immunity. Donor mice were immunized twice with mAb 62/KLH (100 μg/mouse) or mAb 109.3/KLH, as a control, in alum. X-irradiated syngeneic recipients were given reconstituted cell populations (25 × 10⁶ of both T and B) as indicated and then immunized with Tg (10 μg in alum). Assays shown were performed on sera collected at sacrifice, 25 days after the initial challenge with Tg. Results are expressed as mean ± SEM (autoantibodies to Tg) or ± SD (the remaining tests) of individual sera (five to six mice/group). (From Zanetti, M., Glotz, D., and Rogers, J., *J. Immunol.*, 137, 3140, 1986. With permission.)

received cells in various combinations (Figure 5) and were subsequently challenged with Tg. As indicated, T cells from Id-primed mice were indispensable for transferring suppression of the production of autoantibodies irrespective of the type of primed B cells used in association. In converse, Id62-primed B cells, though responsible for the development of intact autologous anti-Id and Id' responses, clearly failed, per se, to transfer suppression. In light of this dichotomy between production of auto-anti-Id and suppression of the response to antigen, we postulate the existence of "regulatory" T cells which, in this instance, are primarily responsible for mediating suppression. The exact phenotype of these cells, i.e., suppressor, Id, or anti-Id as well as the kinetics of their induction, has not been determined yet.

VI. DISCUSSION

Ever since Id were discovered[21,22] and the Id network theory proposed,[1] several aspects emerged which have and continue to modify our understanding of the immune system. New ideas concerning the dynamics of the immune function as well as the teleological function of the immune system have been proposed. Among them, is the hypothesis that the immune system is complete,[8] self-referential,[9] and, whenever self-antigens are involved, self-centered.[10]

The immune response to idiotypic determinants of autologous immunoglobulins is a fundamental aspect of Jerne's network theory.[1] However, it has been argued that, from an operational standpoint, not all Id have regulatory function, i.e., there exists a hierarchy within the immune system and some elements are more important than others.[13] If the hundreds or thousands of different antibodies produced in response to autoantigens were idiotypically unrelated and all independently regulatable, there will be chaos within the immune system and no rules for regulation would be possible. However, the components of the immune system are highly interconnected and regulation through Id is possible in a rather predetermined fashion.[1,23] Thus, some Id are more regulatory than others and, more importantly, some Id could be entirely devoid of regulatory significance. The latter most likely encompass private Id, i.e., those V region determinants practically unique to a given antibody molecule.

The physiological aspect of Id network regulation is better understood through the analysis of the immune response to self-antigens, a process that we better define as the autoimmune network.[10,11] For this reason the analysis presented herein and concerning two fundamental aspects of regulation, namely, regulation through CRI and ReI, is particularly pertinent. Thus, it may be asked, is the distinction between CRI and ReI just a semantic one? Do CRI and ReI apply to the same rule for regulation?

CRI usually account for a sizable fraction of the antibodies produced in response to a given antigen. In most instances, they are quantitatively dominant (reviewed in Reference 24). On the other hand, ReI represent a small fraction of a given antibody response and may not be uniquely specific for the immune response to one antigen only. By virtue of their autoimmunogenic property, they serve as powerful targets of recognition by: (1) anti-Id antibodies and/or (2) T cells which, in turn, regulate B cells and possibly T cells which express more than just ReI receptors. In other terms, ReI serve as specific catalysts of immune reactions. Both definitions were met in the experimental models briefly presented in this communication.

The overall general effect of passive anti-CRI antibodies appears to be suppression. Interestingly, suppression occurred whether the anti-CRI treatment preceded the induction of autoimmunity or intervened during the course of a spontaneous ongoing response. Our experience is shared by the work of other groups.[4,25] Thus, a general consensus would be that CRI are primarily targets for suppression. Is their effect due to direct inactivation of autoreactive clones? The answer is not known. However, in view of the fact that in each of the above-mentioned experiments anti-CRI treatment decreased the total amount of circulating autoantibodies, it is tempting to speculate that other mechanisms might have been involved, including regulation through ReI. In this context, the studies in the Id62 system are particularly enlightening.

At least two aspects emerged from our studies on ReI Id62. The first one is that immunization (vaccination) with anti-Id62 antibodies may, in certain instances, elicit the *de novo* appearance of autoantibodies. We discussed previously[10] and herein the importance of this finding in the context of the autoimmune network concept. Suffice it to say that the mechanism illustrated herein may represent a way through which autoimmunity may be set in motion either by autologous anti-Id or, more importantly, by fortuitously occurring complementary anti-Id produced during the immune response to exogenous, e.g., microbial pathogens, unrelated antigens. Whether this type of autoimmunity can also have pathological consequences is not clear as yet. In our experience, this does not appear to be the case.

A second, and possibly more physiological feature of ReI is, as documented herein in two different experiments, the prevailing suppressive effect of immunity to or by Id62. In both instances, mice suppressed for the total autoantibody response to Tg had, paradoxically, normal or increased serum values of Id62. Therefore, a plausible explanation would be that production of ReI is initially necessary to activate regulatory cells, i.e., those primarily involved in mediating suppression. This contention is supported by the result of adoptive transfer experiments whereby T lymphocytes from Id62-primed animals proved to be required for transferring suppression of the autoantibody response. Thus, it can be concluded that "regulatory" T cells are more important in ReI-dependent regulation. Also, auto-anti-Id may not be directly sufficient for suppression and possibly require clonal interactions within the autoimmune network which involve the participation of "regulatory" T cells.

Are there special structural/genetic features that distinguish CRI from ReI? The structural definition of CRI in antihapten responses is, in most cases, still an open question, although in some systems, it has been defined with some precision.[26] CRI of autoantibodies have been best defined with respect to human rheumatoid factors and appear to correspond to a short amino acid sequence located in the second complementary-determining region of VkIII light chains.[27,28] Collectively, a strong possibility is that CRI originate from germline genes and undergo little, if any, subsequent somatic mutation. On the other hand, ReI of the type alluded to herein also appear to be markers of V_H germline genes which were conserved during diversification processes. This has been elegantly shown by Bona and co-workers[29] for the A48 ReI which was found to be a marker for the V_H 441-4 gene family. With respect to Id62, we recently showed that it can be found in the neonatal spontaneous repertoire of BALB/c mice before any extrauterine antigenic or mitogenic influence and that all Id62-positive autoantibodies belong to the same V_H 7183 gene family.[30] Furthermore, it appears that adult as well as neonatal antibodies bearing Id62 ReI are borne on V_H regions that share identical primary structure.[31] Thus, Id62 ReI may also be the expression of germline genes. An interesting structural feature for Id62 ReI consists of the distribution of the same Id on the heavy and light chains.[32] While the nature of this novel phenomenon has not been elucidated as yet, it is possible that this may constitute one of the structural correlates for ReI.

In conclusion, CRI and ReI both possess the ability to regulate the autoimmune network. A main distinction between them lies in the fact that ReI, unlike CRI, represent only a small fraction of the inducible antibodies. However, ReI are powerful regulators of antiself-responses. Inasmuch as present data favor a germline origin for both of them, differences may be sought in the type of V gene(s) used (are some V genes more important than others?) or in structural characteristics unique for ReI. Although answers to these questions are urgently needed, a further understanding of the mystery of regulation of the autoimmune network requires that more studies similar to the one mentioned herein be made.

ACKNOWLEDGMENTS

The authors with to thank Dr. David H. Katz for his continuous encouragement and Janet Czarnecki for the excellent work in preparing this manuscript.

REFERENCES

1. **Jerne, N. K.,** Towards a network theory of the immune system, *Ann. Immunol. (Paris),* 125, 373, 1974.
2. **Zanetti, M., Rogers, J., and Katz, D. H.,** Induction of autoantibodies to thyroglobulin by anti-idiotypic antibodies, *J. Immunol.,* 133, 240, 1984.
3. **Teitelbaum, D., Rauch, J., Stollar, B. D., and Schwartz, R. S.,** *In vivo* effects of antibodies against a high frequency idiotype of anti-DNA antibodies in MRL mice, *J. Immunol.,* 132, 1282, 1984.
4. **Brown, C. A., Carey, K., and Colvin, R. B.,** Inhibition of auto-immune tubulointerstitial nephritis in guinea pigs by heterologous antisera containing anti-idiotype antibodies, *J. Immunol.,* 123, 2102, 1979.
5. **Zanetti, M. and Bigazzi, P. E.,** Anti-idiotypic immunity and autoimmunity. I. *In vitro* and *in vivo* effects of anti-idiotypic antibodies to spontaneously occurring autoantibodies to rat thyroglobulin, *Eur. J. Immunol.,* 11, 187, 1981.

6. **Zanetti, M., Mampaso, F., and Wilson, C. B.,** Anti-idiotype as a probe in the analysis of autoimmune tubulointerstitial nephritis in the Brown Norway rat, *J. Immunol.,* 131, 1268, 1983.
7. **Katz, D. H.,** Self-recognition as the basis of cell communication in regulation of immune responses, in *Immunopathology, 8th Int. Symp.,* Dixon, F. J. and Miescher, P. A., Eds., Academic Press, New York, 1982, 1.
8. **Coutinho, A., Forni, L., Holmberg, D., Ivars, F., and Vaz, N.,** From an antigen-centered, clonal perspective of immune responses to an organism-centered, network perspective of autonomous activity in a self-referential immune system, *Immunol. Rev.,* 79, 151, 1984.
9. **Jerne, N. K.,** Idiotypic networks and other preconceived ideas, *Immunol. Rev.,* 79, 5, 1984.
10. **Zanetti, M. and Katz, D. H.,** Self recognition, autoimmunity and internal images, in *Current Topics in Microbiology and Immunology,* Vol. 119, Koprowski, H. and Melchers, F., Eds., Springer-Verlag, Heidelberg, 1985, 111.
11. **Zanetti, M.,** The idiotype network in autoimmune processes, *Immunol. Today,* 6, 299, 1985.
12. **Weigle, W. O.,** Analysis of autoimmunity through experimental models of thyroiditis and allergic encephalomyelitis, *Adv. Immunol.,* 30, 159, 1980.
13. **Paul, W. E. and Bona, C.,** Regulatory idiotopes and immune networks: a hypothesis, *Immunol. Today,* 3, 230, 1982.
14. **Zanetti, M.,** Idiotypic regulation of autoantibody production, *Crit. Rev. Immunol.,* 6, 151, 1986.
15. **Bona, C. A., Herber-Katz, E., and Paul, W. E.,** Idiotype-anti-idiotype regulation. I. Immunization with a levan-binding myeloma protein leads to the appearance of auto-anti-(anti-idiotype) antibodies and to the activation of silent clones, *J. Exp. Med.,* 153, 951, 1981.
16. **Zanetti, M.,** Idiotype network and its relevance to autoimmune diseases. Functional considerations, in *Concepts in Immunopathology,* Vol. 3, Cruse, J. M., and Lewis, R. E., Jr., Eds., Karger, S. Basel, 1986, 253.
17. **Zanetti, M. and Wilson, C. B.,** Characterization of anti-tubular basement membrane antibodies in rats, *J. Immunol.,* 130, 2173, 1983.
18. **Zanetti, M., Barton, R. W., and Bigazzi, P. E.,** Anti-idiotypic immunity and autoimmunity. II. Idiotypic determinants of autoantibodies and lymphocytes in spontaneous and experimentally induced autoimmune thyroiditis, *Cell. Immunol.,* 75, 292, 1983.
19. **Zanetti, M., Rogers, J., and Katz, D. H.,** Perturbation of the autoimmune network. I. Immunization with heterologous anti-idiotypic antibodies prior to challenge with antigen induces quantitative variations in the autoantibody response, *Autoimmunity,* in press.
20. **Zanetti, M., Glotz, D., and Rogers, J.,** Perturbation of the auto-immune network. II. Immunization with isologous idiotype induces auto-anti-idiotypic antibodies and suppresses the antibody response elicited by antigen. A serologic and cellular analysis, *J. Immunol.,* 137, 3140, 1986.
21. **Kunkel, H. G., Mannik, M., and Williams, R. C.,** Individual antigenic specificity of isolated antibodies, *Science,* 140, 1218, 1963.
22. **Oudin, J. and Michel, M.,** Une nouvelle forme d'allotypie des globulines du serum de lapin apparemment liee a la fonction et a la specificite anticorps, *C. R. Acad. Sci. (D),* 257, 805, 1963.
23. **Urbain, J., Wuilmart, C., and Cazenave, P.-A.,** Idiotypic regulation in immune networks, in *Contemporary Topics of Molecular Immunology,* Vol. 8, Inman, F. P. and Mandy, W. J., Eds., Plenum Press, New York, 1981, 113.
24. **Bottomly, K.,** All idiotypes are equal but some more equal than others, *Immunol. Rev.,* 79, 45, 1984.
25. **Hahn, B. H. and Ebling, F. M.,** Suppression of murine lupus nephritis by administration of an antiidiotypic antibody to anti-DNA, *J. Immunol.,* 132, 187, 1984.
26. **Capra, J. D. and Fougereau, M.,** One from column A, one from column B, *Immunol. Today,* 4, 175, 1983.
27. **Kunkel, H. G., Winchester, R. J., Joslin, F. G., and Capra, J. D.,** Similarities in the light chains of anti-γ-globulins showing cross-idiotypic specificities, *J. Exp. Med.,* 139, 128, 1974.
28. **Chen, P. P., Goni, F., Fong, S., Jirik, F., Vaughan, J. H., Frangione, B., and Carson, D. A.,** The majority of human monoclonal IgM rheumatoid factors express a "primary structure-dependent" cross-reactive idiotype, *J. Immunol.,* 134, 3281, 1985.
29. **Victor-Kobrin, C., Bonilla, F., Bellon, B., and Bona, C. A.,** Immunochemical and molecular characterization of regulatory idiotypes expressed by monoclonal antibodies exhibiting or lacking β2-6 fructosan binding activity, *J. Exp. Med.,* 162, 647, 1985.
30. **Glotz, D. and Zanetti, M.,** Detection of a regulatory idiotype on a spontaneous neonatal self-reactive hybridoma antibody, *J. Immunol.,* 137, 223, 1986.
31. **Haseman, C. A., Meek, K. D., Glotz, D., Sollazzo, M., Zanetti, M., and Capra, D.,** Molecular characterization of murine anti-thyroglobulin antibody molecules: a germline V_H auto-antibody response, submitted for publication.
32. **Zanetti, M., Liu, F.-T., Rogers, J., and Katz, D. H.,** Heavy and light chains of a mouse monoclonal autoantibody express the same idiotype, *J. Immunol.,* 135, 1245, 1985.

Chapter 8

ANTI-IDIOTYPIC ANTIBODIES IN CANCER THERAPY

D. Herlyn and H. Koprowski

TABLE OF CONTENTS

I. INTRODUCTION

The humoral and cellular immune responses to a growing tumor have been studied extensively in cancer patients (reviewed in References 1 to 4); however, the immunoregulatory functions of these responses are not well characterized. In general, cancer patients' sera have been shown to contain antibodies that bind to both tumor and normal tissue.[1,5] When lymphocytes obtained from cancer patients were either transformed by Epstein-Barr virus (EBV) or fused with mouse myelomas to produce interspecies hybridomas, antibodies were secreted that in a few cases were restricted in their binding specificity to tumors of the same tissue origin as the patient's tumor, but in most cases[6-10] bound also to tumors of different origin.[6-8,11-17] In general, these antibodies did not bind to any of the normal tissues tested.[6-8,10,13,14] However, the clinical significance of these responses remains unclear, since most of those studies did not include lymphocytes or normal sera from healthy individuals. Although antibodies that are reactive with the tumor cell surface and may, thus, be functionally important have been described,[5-16,18-26] those that bind exclusively to intracytoplasmic antigens seem to predominate;[13,16,17] of course, such antigens may not display any biological functions.

Although a few attempts have been made to correlate the presence of antitumor antibody with prolonged survival of some cancer patients,[27-29] those studies were without reference to an adequate number of control patients who did not develop antibodies. Demonstrations of complement-mediated lysis of astrocytomas,[26] melanomas,[29] and acute myeloid leukemias[9] by polyclonal, monospecific, and monoclonal human antibodies, respectively, are isolated findings that do not allow any definite conclusion about the role, if any, of anticancer antibody in human malignancy.

It is impossible at present to determine whether the vast majority of cancer patients are incapable of mounting an immune response or whether the response remains undetected. One can speculate that cancer patients are capable of producing cancer antibodies, but that the antibodies "disappear" from the circulation through binding to anti-idiotypic antibodies (anti-Id or Ab_2). Alternatively, an immune response against the patient's own tumor may have been elicited by induction of specific B cells that are primed by antigen, but incapable of antibody secretion. Until now, no tools were available that could be used to investigate either of these hypotheses. The development of Ab_2 bearing the internal image of a tumor antigen[30] (see below) now provides this tool. Such Ab_2 may be used to detect those T and B cells of the patient that are produced as the result of the patient's immune response to his own tumor. Alternatively, active immunization of a patient with Ab_2 may induce an immune response characterized by production of anti-anti-Id antibodies (Ab_3), which, in turn, may recognize tumor antigens and even bring about destruction of the tumor.

II. ROLE OF Ab_2 INDUCED IN CANCER PATIENTS DURING IMMUNOTHERAPY WITH MONOCLONAL ANTITUMOR ANTIBODY

We have shown that gastrointestinal cancer patients who were treated with monoclonal antibody (MAb) 17-1A improved clinically and had long remission from their diseases.[31] Although the MAb may have directly destroyed tumor cells by interaction with macrophages or killer cells,[32,33] other mechanisms may have been operative, especially since antitumor effects were often detected late (>3 months) after MAb treatment. In some patients, these effects were measurable over a period of several months following a single injection of MAb[31] In these patients, MAb 17-1A circulated in the blood for less than 20 days and MAb binding to metastatic tumor biopsy samples was detected only for up to 1 week after the injection of the MAb.[31,34] As an alternative mechanism of antitumor effects, we have postulated a network of interacting anti-idiotypic T and B cells directed against the MAb. We

Table 1
CHARACTERISTICS OF HUMAN Ab$_2$ TO
ANTICOLORECTAL CARCINOMA MAb 17-1A

Parameter investigated	Results[a]	Total No. of patients
No. of patients with Ab$_2$ responses	35	41
Ab$_2$ (μg/mℓ) isolated from patients' sera (range in various patients)	2.8 — 42.0	4
Ab$_2$ as % of total antimouse IgG (range)	21.0 — 80.2	4
Anticombining site Ab$_2$[b]		
Maximal % inhibition of ^{125}I-17-1A binding to CRC cells by Ab$_2$ (range)	32.7 — 70.5	4
Maximal % inhibition of ^{125}I-17-1A binding to human Ab$_2$ by extracts from CRC cells (range)	20.0 — 68.5	3
Cross reactivities between various human Ab$_2$; maximal % inhibition of ^{125}I-Ab$_2$ (from one patient) binding to MAb 17-1A by Ab$_2$ derived from various other patients (range)	17.0 — 91.2	4

[a] See References 35 and 36 for details.
[b] Ab$_2$ to the antigen-combining site of Ab$_1$.

could show that patients who improved clinically had developed high titers of Ab$_2$ and that these titers were maintained for prolonged time periods.[35,36] The characteristics of the Ab$_2$ are summarized in Table 1.

Eighty-four percent of the patients who developed antimouse IgG antibody also developed Ab$_2$. Between 2.8 and 42.0 μg of Ab$_2$ could be isolated per milliliter serum in various patients, and Ab$_2$ comprised between 21 and 80% of the total human antimouse IgG antibody response. In assays for the presence of anticombining site Ab$_2$, binding of ^{125}I-17-1A MAb to CRC cells was inhibited by Ab$_2$ derived from 4 patients between 32 and 70% and binding of ^{125}I-17-1A to Ab$_2$ from 3 patients was inhibited by CRC cell extracts between 20 and 68%. Thus, a significant proportion of the Ab$_2$ bound to the antigen-combining site of MAb 17-1A. The Ab$_2$ isolated from sera of four patients showed significant cross reactivities, indicating that the patients' Ab$_2$ recognize a common determinant(s) on the MAb.[35,36] The presence of Ab$_2$ was further confirmed by its production in tissue culture by lymphocytes stimulated in vitro with MAb 17-1A.[35] In light of these results, we hypothesized that the Ab$_2$ produced by the patients would elicit the formation of Ab$_3$ reacting with both Ab$_2$ and tumor cells. Although no antitumor antibody could be found in the patients' sera, presumably because it was complexed to Ab$_2$ circulating in large amounts,[35,36] such antibodies were produced by patients' lymphocytes in culture following stimulation with autologous Ab$_2$. Furthermore, in the course of 4 immunizations with human Ab$_2$ of a patient who had developed Ab$_2$ following the injection of MAb 17-1A 2 years before, production of Ab$_2$ and tumor-binding antibody by the patient's lymphocytes was observed following their stimulation in vitro by either Ab$_1$ or Ab$_2$.[37] However, the specificity of the Ab$_2$ effects could not be studied in detail because of the small amounts of antibody produced in culture and the inavailability of 17-1A tumor antigen.*

---

* Since the completion of those studies, the antigen defined by the monoclonal Ab$_1$ 17-1A has been identified[39] and isolated.[40]

Table 2
INDUCTION OF IMMUNITY TO TUMORS BY Ab₂

Origin of tumor	Ab₁			Ab₂			Immunity induced by Ab₂				Ref.
	Induced in	Clonality	Binding specificity	Induced in	Clonality	Binding specificity	Induced in	Humoral/cell-mediated	Specificity	Effect on challenge with tumor	
Human adenocarcinoma of gastrointestinal tract	BALB/c mouse	Monoclonal	30,000-Dalton protein on human gastrointestinal carcinoma-associated antigen	Goat	Polyclonal	Combining site-associated Id on homologous Ab₁ (no other Ab₁ with same epitope specificity available)	BALB/c mouse Rabbit	Ab₃ (monoclonal) Ab₃ (polyclonal)	30,000-Dalton protein on human gastrointestinal carcinomas	n.t. because of lack of suitable animal model system	30
Human melanoma	BALB/c mouse	Monoclonal	97,000-Dalton melanoma-associated antigen	Rabbit	Polyclonal	Combining site-associated cross-reactive Id on several anti-p97Ab from BALB/c mice with same epitope specificity as Ab₁	BALB/c mouse	Ab₃ DTH	97,000-Dalton protein on human melanoma / Melanomas with high antigen expression only (>380,000 antigenic sites per cell)	n.t. because of lack of suitable animal model system	42
SV40-induced murine tumor	BALB/c mouse	Monoclonal	SV40 T-Ag	Rabbit	Polyclonal	Combining site-associated private Id on homologous Ab₁ only	BALB/c mouse	Ab₃	Bound to Ab₃ only, but not to antigen	Partial inhibition of SV40-transformed tumor growth in vivo	43
Murine mammary tumor	BALB/c mouse	Monoclonal	gp52 of MMTV	BALB/c mouse	Monoclonal	Combining site-associated Id on Ab₁	DBA/2 mouse	DTH	n.t.	Prolongation of survival following challenge with lymphoma	44
Murine sarcoma	Not available			BALB/c[a] mouse	Monoclonal	n.t. because of inavailability of Ab₁	BALB/c	DTH	DTH induced to sarcoma used to immunize mice for Ab₂, not to unrelated sarcomas	n.t.	46
Murine transitional cell bladder carcinoma	Rat	Monoclonal	175,000-Dalton protein on BALB/c mouse transitional cell bladder carcinomas	BALB/c mouse	Monoclonal	Combining site-associated Id on Ab₁	BALB/c	DTH	Antigen-positive bladder carcinomas only	n.t.	45

Note: Abbreviations: DTH, delayed-type hypersensitivity; LAI, leukocyte adherence inhibition; MMTV, mouse mammary tumor virus; n.t., not tested. SV40 T-Ag, Simian virus 40 T-antigen.

a   Ab₂ was isolated from BALB/c mice hyperimmunized with syngeneic sarcoma cells.

III. INDUCTION OF IMMUNITY TO HUMAN TUMORS BY Ab₂ IMMUNIZATIONS

These studies are summarized in Table 2. We have produced Ab_2 of the internal antigen image type that may have potential as an immunizing agent against human cancer. Antihuman tumor MAb GA733 (Ab_1) was chosen for production of Ab_2 based on its binding specificity for human carcinomas of various tissue origins and tumoricidal activity both in vitro and in vivo in nude mice.[38] In addition, the tumor antigen defined by MAb GA733 has been isolated, extensively characterized biochemically, and amino acid residues of the protein partially sequenced.[39] Ab_2 against Ab_1 GA733 was produced in goats and extensively characterized in in vitro binding and binding inhibition studies and in vivo following its injection into mice and rabbits. The results[30] can be summarized as follows: (1) the Ab_2 inhibited binding of Ab_1 to target cells in culture, indicating that the Ab_2 was directed to the antigen-combining site of Ab_1; (2) binding inhibition of Ab_1 to Ab_2 by tumor antigen was maximally 67%; thus, approximately 67% of the Ab_2 molecules were combining site specific; and (3) in two different species of animals, the Ab_2 induced the formation of Ab_3 in the absence of antigen. Monoclonal Ab_3 were derived from BALB/c mice and polyclonal Ab_3 from rabbits. Both the murine and the rabbit-derived Ab_3 immunoprecipitated the same molecular weight fraction from isolated GA733 antigen as Ab_1 (Figure 1) and exhibited the same binding specificity to various target cells as Ab_1 (Figure 2). These results suggested that the Ab_2 bears the internal image of the GA733 tumor antigen. The Ab_3 was of IgG1 isotype and, as such, not tumoricidal, since antitumor effects have been observed only with IgG2a Ab_1 that bind to the Fc receptors on either murine[33,41] or human macrophages.[32]

The ultimate aim of Ab_2 immunizations is to induce protective immunity against tumors. Since human tumors can only be grown in immunodeficient animals, the possible immunotherapeutic effects of Ab_2 immunizations must ultimately be evaluated directly in patients. The Ab_2 bearing the internal image of GA733 tumor antigen may have great potential for modulating the immune responses of cancer patients to their tumors.

Nepom et al.[42] have produced Ab_2 in rabbits to murine MAb against melanoma-associated antigen p97 (Ab_1). The Ab_2 recognized combining site-associated idiotopes on Ab_1 that were shared by other anti-p97 MAb with the same epitope specificity as Ab_1. Mice immunized with the Ab_2 demonstrated delayed-type hypersensitivity (DTH) reactions when challenged with melanoma cells expressing 380,000 p97 molecules per cell, but not when challenged with melanoma cells expressing only 2600 such molecules. Ab_2-immunized mice also developed antibodies (Ab_3) that inhibited binding of Ab_2 to Ab_1 and precipitated p97 antigen from radiolabeled melanoma cell lysates, whereas none of the control sera showed this reactivity. It is unclear whether the Ab_2 used in that study were of the internal antigen image type, since all of the Ab_1 and the Ab_3 involved were derived from the same strain of mice (BALB/c). The demonstration that Ab_2 reacts with Ab_1 of various species directed against the same epitope on the p97 antigen and/or the induction of Ab_3 in a species different from that used for generation of Ab_1 might clarify this point. Furthermore, the immunotherapeutic potential of the Ab_3 induced is questionable, since binding of the Ab_3-containing sera to the surface of melanoma cells has not been demonstrated.

IV. Ab₂ INDUCES IMMUNITY IN MICE TO MURINE TUMORS

These studies are summarized, also, in Table 2. Kennedy et al.[43] have induced protective immunity in mice against Simian virus 40 (SV40)-induced tumors by Ab_2 immunizations. The Ab_2 were produced in rabbits against MAb to SV40 T antigen (Ab_1). The Ab_2 were directed against "private" Id associated with the combining site of the homologous Ab_1 and were not expressed by any of the other Ab_1 against the SV40 T antigen. It is not known

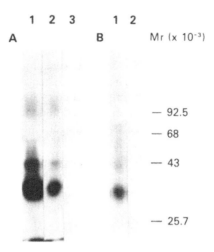

FIGURE 1.   Panel A: Electrophoretic profile of ^{125}I-labeled GA733 antigen after immunoprecipitation with purified Ab$_1$ GA733 (lane 1), murine monoclonal Ab$_2$ CE5 elicited by goat Ab$_2$ to Ab$_1$ GA733 (lane 2), and unrelated MAb ME7529 (lane 3). The major band at 30,000 daltons and the minor band at 40,000 daltons represent the ^{125}I-labeled GA733 antigen. All antibodies were used at a concentration of 2.7 μg/mℓ. Panel B: Electrophoretic profile of ^{125}I-labeled GA733 antigen bound to rabbit Ab$_3$ isolated from serum by adsorption to glutaraldehyde-fixed SW948 CRC cells and elution of bound antibody with glycine-HCl buffer, pH 2.8 (lane 1); absence of binding to normal rabbit IgG (lane 2).[30] (From Herlyn, D., Ross, A. H., and Koprowski, H., *Science*, 232, 100, 1986. Copyright 1986 by the AAAS. With permission.)

whether the Ab$_2$ were of the internal antigen image type, since the exact epitope specificity of the anti-T-antigen Ab$_1$ has not been described. Moreover, the mechanism underlying the observed tumor growth inhibition induced by Ab$_2$ immunizations is unclear, since neither antigen-binding Ab$_3$ nor cytolytic T cells were induced in the protected mice.

Raychaudhuri et al.[44] reported on the inhibition of murine lymphoma growth by Ab$_2$ immunizations. In that study, monoclonal BALB/c mouse-derived Ab$_2$ against the combining site of a MAb to the envelope glycoprotein gp52 of the mammary tumor virus were coupled to keyhole limpet hemocyanin and injected into DBA/2 mice. The immunized animals showed DTH reaction against a challenge with irradiated lymphoma cells. Mice immunized with a mixture of three monoclonal Ab$_2$ and challenged with live lymphoma cells survived, whereas unvaccinated tumor-bearing mice died. Presumably, the Ab$_2$ used in that study bear internal images of the gp52 tumor-associated antigen, based on the anticombining site specificity of the Ab$_2$, although additional data such as the reactivity of the Ab$_2$ with Ab$_1$ from various species that bind to the same epitope on the gp52 antigen and/or the elicitation of antigen-specific immune responses across species barriers would be required to confirm the nature of the Ab$_2$.

DTH to a syngeneic murine bladder carcinoma[45] and sarcoma[46] has been induced by Ab$_2$ immunizations. In the first case, Ab$_1$ against a 175,000-dalton protein antigen on BALB/c mouse transitional cell bladder carcinoma was used to elicit monoclonal Ab$_2$; in the sarcoma system, monoclonal Ab$_2$ was produced in mice hyperimmunized with syngeneic sarcoma cells. Due to the lack of corresponding Ab$_1$ in the latter system, the binding specificity of the Ab$_2$ could not be investigated. In both systems, DTH induced with Ab$_2$ was tumor specific. In the sarcoma system,[46] induction of DTH was allotype restricted, i.e., required homology at the Igh-1 locus between host and Ab$_2$.

V. APPROACHES TO ANTITUMOR THERAPY WITH Ab$_2$

In approaches to the development of Ab$_2$ as an immunotherapeutic agent against cancer, the choice of both Ab$_1$ and Ab$_2$ are most critical.

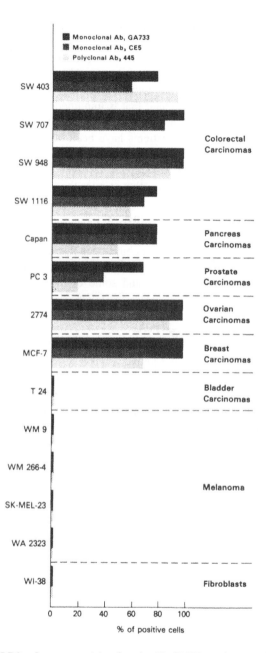

FIGURE 2. Immunoreactivity of murine Ab₁ GA733, murine monoclonal Ab₃ CE5, and rabbit polyclonal Ab₃ 445. Percentage of antibody-reactive cells was determined by mixed hemadsorption assay.[30] (From Herlyn, D., Ross, A. H., and Koprowski, H., *Science*, 232, 100, 1986. Copyright 1986 by the AAAS. With permission.)

A. Antitumor Antibodies (Ab₁) for Production of Ab₂

1. The Ab₁ should exhibit preferential binding reactivity to tumor cells as compared to normal cells. Although absolute tumor specificity for antigens most likely does not exist, many tumor-associated antigens are expressed at significantly higher densities on tumor cells than on normal cells.[47] Sufficient antigen density is an important

determinant of tumor destruction by antibodies.[48] Monoclonal antigastrointestinal carcinoma antibody 17-1A, with high binding reactivity to tumors of the gastrointestinal tract and low binding reactivity to normal colonic mucosa,[34] has been administered to cancer patients without producing any adverse side effects related to the binding reactivity of the antibody to normal colonic mucosa.[31,49]

2. The Ab_1 should bind to a determinant that is immunodominant for most individuals. It is very difficult to define such determinants, since antibody responses to defined human tumor antigens in cancer patients have been observed only in isolated cases (see above).

3. The Ab_1 should mediate cytotoxic reactivities to tumor cells. Ab_2 against tumoricidal Ab_1 may induce the formation of Ab_3 that mimics the binding specificity of Ab_1 and exhibits tumoricidal activity. However, effector cell-mediated cytotoxic activity will be dependent on the isotype of the elicited Ab_3.[41]

4. The Ab_1 should bind to an antigen that is available in purified form. Several antigens expressed by a variety of human tumors have been purified and amino acid sequences of some protein antigens are available.[50] Carbohydrate determinants of glycolipid antigens have been characterized[51,52] and some of those are available in synthetic form.[59]

B. Ab₂ of the Internal Antigen Image Type

This Ab_2 should bind to various Ab_1 derived from different hosts and binding to the same epitope. This presents difficulties in the human cancer system, since only a few Ab_1 are available that fulfill these criteria. In the case of Ab_1 associated with colorectal cancer, three MAb have been generated after immunization of mice with different tumor preparations and the Ab_2 developed against one of the Ab_1 bound to all three Ab_1.[40] The binding of Ab_2 of the internal antigen image type to Ab_1 should be significantly inhibited by tumor cell extracts or preferably by purified cancer antigen. This would indicate that Ab_2 is directed against the antigen-combining site of Ab_1. However, these in vitro characteristics of Ab_2 may not distinguish unequivocally internal antigen image-bearing Ab_2 from the Ab_2 directed to cross-reactive idiotopes (CRI, see below).

Ultimately, the internal image nature of the selected Ab_2 will be demonstrated in in vivo immunizations of animals of various species to produce antigen-specific Ab_3 and/or anti-anti-idiotypic cellular immune responses. The cellular responses can be evaluated in tissue culture by stimulation of the cultured cells with tumor antigen and/or Ab_2. The Ab_3 present in serum should bind specifically to the Ab_2 and, most importantly, to the cancer cell or purified cancer antigen. Genetic restriction of the host's immune responses to immunization with Ab_2 of the internal image type is not greater than that to the original antigen. Thus, to demonstrate the internal image nature of Ab_2, animals from at least two different species should be immunized with Ab_2 for production of Ab_3.

Since induction of an immune response by Ab_2 has been shown to be dose dependent,[53,54] it is important to immunize with various dosages of Ab_2 in order to induce a maximum immune response in the host. The outcome of immunizations with Ab_2 may also depend on the state of fragmentation of Ab_2 (intact antibody or F [ab']$_2$ fragments) as shown in several instances.[55]

C. Anti-CRI Ab₂

CRI have been described in antibody populations derived from different individuals of the same species, or even from individuals belonging to different species in response to the same antigen (reviewed in Reference 56). Thus, expression of a particular CRI by antibodies against a given antigen is often closely associated with the binding specificity of these antibodies. However, since the same CRI might also be expressed on antibodies directed to unrelated antigen(s), Ab$_2$ against CRI may, by stimulating CRI-positive B cells, elicit various Ab$_3$ populations that share the same CRI, but differ in their antigen-binding specificities. Bona[57] has characterized the CRI expressed by antibodies with various binding specificities as regulatory idiotopes.

Immune responses induced by anti-CRI Ab$_2$ are restricted to those species that are genetically capable of expressing the particular CRI. Thus, immunization with anti-CRI Ab$_2$ is more restricted genetically than immunization with internal image Ab$_2$. The proportion of lymphocytes and Ab$_3$ that specifically binds antigen will be much lower when anti-CRI Ab$_2$ are used for immunization as compared to immunizations with internal image Ab$_2$. Finally, and perhaps most importantly, anti-CRI Ab$_2$ might induce immune responses of unwanted specificities, i.e., those directed to antigens expressed not only by cancer cells, but also by normal cells.

VI. ADVANTAGES OF Ab$_2$ AS IMMUNOTHERAPEUTIC AGENTS AGAINST CANCER

Ab$_2$ of the internal antigen image type offer numerous advantages over purified tumor antigens for treatment of cancer patients:

1. Ease and economy of production. Ab$_2$ can be easily prepared by the hybridoma technique or from sera of immune animals, whereas isolation of purified tumor antigens in large amounts is extremely difficult. Thus, immunizations with Ab$_2$, particularly those with Ab$_2$ from hybridomas, represent the least costly regimen of cancer therapy.
2. Safety. Ab$_2$ preparations, unlike tumor antigen preparations, do not involve any tumor cell components, thus, circumventing the cumbersome search for traces of tumor-derived nucleic acids in products that are ultimately designed for human use.
3. Specificity. Ab$_2$, in particular monoclonal Ab$_2$, exhibit highly restricted specificity by virtue of mimicry of a single tumor antigen epitope, and the use of either F(ab')$_2$ or Fab fragments of Ab$_2$ for immunizations would minimize the induction of antiisotypic and/or -allotypic immune responses. Conversely, isolated tumor antigens may contain numerous determinants shared with normal host tissues and might, thus, induce a deleterious autoimmune response in the host. Although the use of synthetic peptides or products obtained by gene cloning may theoretically provide more defined compounds, these products sometimes lack the conformational structure of the original antigen and are, thus, able to elicit a specific immune response.
4. Breakage of immunological tolerance. Ab$_2$ have been shown to induce immune responses in neonatal hosts who were incapable of responding directly to antigen. For instance, priming of neonatal mice with Ab$_2$ to antibodies directed against capsular antigens of pathogenic bacteria (*Escherichia coli*), followed by booster immunization with bacterial antigen, led to protection of the mice against a lethal challenge with live bacteria. In contrast, priming of mice with bacterial antigen alone did not induce protective immunity.[58] Thus, Ab$_2$ can stimulate B-cell clones that are tolerant to stimulation by antigen, suggesting that usefulness of Ab$_2$ in inducing immunity against tumor antigens to which a state of tolerance often exists.

REFERENCES

1. **Old, L. J.,** Cancer immunology: the search for specificity, *Cancer Res.,* 41, 361, 1981.
2. **Hirshaut, Y. and Slovin, S. F.,** Harnessing T-lymphocytes for human cancer immunotherapy, *Cancer,* 56, 1366, 1985.

3. **Moore, M.,** Natural immunity to tumors — theoretical predictions and biological observations, *Br. J. Cancer,* 52, 147, 1985.

4. **Olsson, L.,** Human monoclonal antibodies in experimental cancer research, *J. Natl. Cancer Inst.,* 75, 397, 1985.

5. **Freundschuh, M., Shiku, H., Takahashi, T., Ueda, R., Ransohoff, J., Oettgen, H. F., and Old, L. J.,** Serological analysis of cell surface antigens of malignant human brain tumors, *Proc. Natl. Acad. Sci. U.S.A.,* 75, 5122, 1978.

6. **Sikora, K., Alderson, T., Phillips, J., and Watson, J. V.,** Human hybridomas from malignant gliomas, *Lancet,* i, 11, 1982.

7. **Glassy, M. C., Handley, H. H., Hagiwara, H., and Royston, I.,** UC 729-6, a human lymphoblastoid B-cell line useful for generating antibody-secreting human-human hybridomas, *Proc. Natl. Acad. Sci. U.S.A.,* 80, 6327, 1983.

8. **Watson, D. B., Burns, G. F., and Mackay, I. R.,** In vitro growth of B lymphocytes infiltrating human melanoma tissue by transformation with EBV: evidence for secretion of anti-melanoma antibodies by some transformed cells, *J. Immunol.,* 130, 2442, 1983.

9. **Olsson, L., Andreasen, R. B., Ost, A., Christensen, B., and Biberfeld, P.,** Antibody producing human-human hybridomas. II. Derivation and characterization of an antibody specific for human leukemic cells, *J. Exp. Med.,* 159, 537, 1984.

10. **Murakami, H. J., Hashizume, S., Ohashi, H., Shinohara, K., Yasumoto, K., Nomoto, K., and Omura, H.,** Human-human hybridomas secreting antibodies specific to human lung carcinoma, *In Vitro Cell, Dev. Biol.,* 21, 593, 1985.

11. **Cahan, D. L., Irie, R. F., Singh, R., Cassidenti, A., and Paulson, J. C.,** Identification of a human neuroectodermal tumor antigen (OFA-I-2) as ganglioside GD2, *Proc. Natl. Acad. Sci. U.S.A.,* 79, 7629, 1982.

12. **Irie, R. F., Sze, L. L., and Saxton, R. E.,** Human antibody to OFA-I, a tumor antigen, produced in vitro by Epstein-Barr virus-transformed human B-lymphoid cell lines, *Proc. Natl. Acad. Sci. U.S.A.,* 79, 5666, 1982.

13. **Houghton, A. N., Brooks, H., Cote, R. J., Taormina, M. C., Oettgen, H. F., and Old, L. J.,** Detection of cell surface and intracellular antigens by human monoclonal antibodies. Hybrid cell lines derived from lymphocytes of patients with malignant melanoma, *J. Exp. Med.,* 158, 53, 1983.

14. **Sikora, K., Alderson, T., Ellis, J., Phillips, J., and Watson, J.,** Human hybridomas from patients with malignant disease, *Br. J. Cancer,* 47, 135, 1983.

15. **Tai, T., Paulsen, J. C., Cahan, L. D., and Irie, R. F.,** Ganglioside GM2 as a human tumor antigen (OFA-I-1), *Proc. Natl. Acad. Sci. U.S.A.,* 80, 5392, 1983.

16. **Cote, R. J., Morrissey, D. M., Oettgen, H. F., and Old, L. J.,** Analysis of human monoclonal antibodies derived from lymphocytes of patients with cancer, *Fed. Proc. Fed. Am. Soc. Exp. Biol.,* 43, 2465, 1984.

17. **Imam, A., Drushella, M. M., Taylor, C. R., and Tokes, Z. A.,** Generation and immunohistological characterization of human monoclonal antibodies to mammary carcinoma cells, *Cancer Res.,* 45, 263, 1985.

18. **Carey, T. E., Takahashi, T., Resnick, L. A., Oettgen, H. F., and Old, L. J.,** Cell surface antigen of human malignant melanoma: mixed hemadsorption assays for humoral immunity to cultured autologous melanoma cells, *Proc. Natl. Acad. Sci. U.S.A.,* 73, 3278, 1976.

19. **Shiku, H., Takahashi, T., Oettgen, H. F., and Old, L. T.,** Cell surface antigens of human malignant melanoma. II. Serological typing with immune adherence assays and definition of two new surface antigens, *J. Exp. Med.,* 144, 873, 1976.

20. **Shiku, H., Takahashi, T., Resnick, L. A., Oettgen, H. F., and Old, L. J.,** Cell surface antigens of human malignant melanoma. III. Recognition of autoantibodies with unusual characteristics, *J. Exp. Med.,* 145, 784, 1977.

21. **Real, F. X., Mattes, M. J., Houghton, A. N., Oettgen, H. F., Lloyd, K. O., and Old, L. J.,** Class I (unique) tumor antigens of human melanoma. Identification of a 90,000 dalton cell surface glycoprotein by autologous antibody, *J. Exp. Med.,* 160, 1219, 1984.

22. **Sikora, K. and Wright, R.,** Human monoclonal antibodies to lung-cancer antigens, *Br. J. Cancer,* 43, 696, 1981.

23. **Cole, S. P. C., Campling, B. G., Louwman, I. H., Kozbor, D., and Roder, J. C.,** A strategy for the production of human monoclonal antibodies reactive with lung tumor cell lines, *Cancer Res.,* 44, 2750, 1984.

24. **Vlock, D. R. and Kirkwood, J. M.,** Serial studies of autologous antibody reacting to melanoma. Relationship to clinical course and circulating immune complexes, *J. Clin. Invest.,* 76, 849, 1985.

25. **Dent, P. B., Liao, S.-K., McCulloch, P. B., Stone, B. R., and Singal, D. P.,** Absence of melanoma specificity in the reactivity of melanoma patients' sera with cultured allogeneic melanoma cell lines, *Cancer,* 49, 2043, 1982.

26. **Phillips, J. P., Sujatanoud, M., Martuza, R. L., Quindlen, E. A., Wood, W. C., Kornblith, P. L., and Dohan, F. C., Jr.,** Cytotoxic antibodies in preoperative glioma patients: a diagnostic assay, *Acta Neurochir.*, 35, 43, 1976.
27. **Sikora, K. and Phillips, J.,** Human monoclonal antibodies to glioma cells, *Br. J. Cancer,* 43, 105, 1981.
28. **Jones, P. C., Sze, L. L., Liu, P. Y., Morton, D. L., and Irie, R. F.,** Prolonged survival for melanoma patients with elevated IgM antibody to oncofetal antigen, *J. Natl. Cancer Inst.,* 66, 249, 1981.
29. **Irie, R. F., Jones, P. C., Morton, D. L., and Sidell, N.,** In vitro production of human antibody to a tumour-associated foetal antigen, *Br. J. Cancer,* 44, 262, 1981.
30. **Herlyn, D., Ross, A. H., and Koprowski, H.,** Anti-idiotypic antibodies bear the internal image of a human tumor antigen, *Science,* 232, 100, 1986.
31. **Sears, H. F., Herlyn, D., Steplewski, Z., and Koprowski, H.,** Effects of monoclonal antibody immunotherapy on patients with gastrointestinal adenocarcinoma, *J. Biol. Response Modifiers,* 3, 138, 1984.
32. **Steplewski, Z., Lubeck, M., and Koprowski, H.,** Human macrophages armed with murine immunoglobulin G2a antibodies to tumors destroy human cancer cells, *Science,* 221, 865, 1983.
33. **Herlyn, D., Herlyn, M., Steplewski, Z., and Koprowski, H.,** Monoclonal anti-human tumor antibodies of six isotypes in cytotoxic reactions with human and murine effector cells, *Cell. Immunol.,* 92, 105, 1985.
34. **Shen, J. W., Atkinson, B., Koprowski, H., and Sears, H. F.,** Binding of murine immunoglobulin to human tissues after immunotherapy with anticolorectal carcinoma monoclonal antibody, *Int. J. Cancer,* 33, 465, 1984.
35. **Koprowski, H., Herlyn, D., Lubeck, M., DeFreitas, E., and Sears, H. F.,** Human anti-idiotype antibodies in cancer patients: is the modulation of the immune response beneficial for the patient?, *Proc. Natl. Acad. Sci. U.S.A.,* 81, 216, 1984.
36. **Herlyn, D., Lubeck, M., Sears, H. F., and Koprowski, H.,** Specific detection of anti-idiotypic immune responses in cancer patients treated with murine monoclonal antibody, *J. Immunol. Methods,* 85, 27, 1985.
37. **DeFreitas, E., Suzuki, H., Herlyn, D., Lubeck, M., Sears, H., Herlyn, M., and Koprowski, H.,** Human antibody induction to the idiotypic and anti-idiotypic determinants of a monoclonal antibody against a gastrointestinal carcinoma antigen, in *Current Topics in Microbiology and Immunology,* Vol. 119, Melchers, F., Ed., Springer-Verlag, Heidelberg, 1985, 76.
38. **Herlyn, D., Herlyn, M., Ross, A. H., Ernst, C., Atkinson, B., and Koprowski, H.,** Efficient selection of human tumor growth-inhibiting monoclonal antibodies, *J. Immunol. Methods,* 73, 157, 1984.
39. **Ross, A. H., Herlyn, D., Iliopoulos, D., and Koprowski, H.,** Isolation and characterization of a carcinoma-associated antigen, *Biochem. Biophys. Res. Commun.,* 135, 297, 1986.
40. **Gottlinger, H. G., Funke, I., Johnson, J. P., Gokel, J. M., and Riethmuller, G.,** The epithelial cell surface antigen 17-1A, a target for antibody-mediated tumor therapy: its biochemical nature, tissue distribution and recognition by different monoclonal antibodies, *Int. J. Cancer,* 38, 47, 1986.
41. **Herlyn, D. and Koprowski, H.,** IgG2a monoclonal antibodies inhibit human tumor growth through interaction with effector cells, *Proc. Natl. Acad. Sci. U.S.A.,* 79, 4761, 1982.
42. **Nepom, G. T., Nelson, K. A., Holbeck, S. L., Hellstrom, I., and Hellstrom, K. E.,** Induction of immunity to human tumor marker by *in vivo* administration of anti-idiotypic antibodies in mice, *Proc. Natl. Acad. Sci. U.S.A.,* 81, 2864, 1984.
43. **Kennedy, R. C., Dreesman, G. R., Butel, J. S., and Lanford, R. E.,** Suppression of in vivo tumor formation induced by simian virus 40-transformed cells in mice receiving antiidiotypic antibodies, *J. Exp. Med.,* 161, 1432, 1985.
44. **Raychaudhuri, S., Fuji, H., and Kohler, H.,** Idiotype vaccines against tumors, Abstract, Symp. 2nd S-W Foundation for Biomedical Research Virology/Immunology, San Antonio, Tex., December 4 to 6, 1985.
45. **Lee, V. K., Harriott, T. G., Kuchroo, V. K., Halliday, W. J., Hellstrom, I., and Hellstrom, K. E.,** Monoclonal anti-idiotypic antibodies related to a murine oncofetal bladder tumor antigen induce specific cell-mediated tumor immunity, *Proc. Natl. Acad. Sci. U.S.A.,* 82, 6286, 1985.
46. **Forstrom, J. W., Nelson, K. A., Nepom, G. T., Hellstrom, I., and Hellstrom, K. E.,** Immunization to a syngeneic sarcoma by a monoclonal auto-anti-idiotypic antibody, *Nature,* 303, 627, 1983.
47. **Herlyn, M., Thurin, J., Balaban, G., Bennicelli, J. L., Herlyn, D., Elder, D. E., Bondi, E., Guerry, D., Nowell, P. C., Clark, W. H., and Koprowski, H.,** Characteristics of cultured human melanocytes isolated from different stages of tumor progression, *Cancer Res.,* 45, 5670, 1985.
48. **Herlyn, D., Powe, J., Ross, A. H., Herlyn, M., and Koprowski, H.,** Inhibition of human tumor growth by IgG2a monoclonal antibodies correlates with antibody density on tumor cells, *J. Immunol.,* 134, 1300, 1985.
49. **Sears, H. F., Mattis, J., Herlyn, D., Hayry, P., Atkinson, B., Ernst, C., Steplewski, Z., and Koprowski, H.,** Phase I clinical trial of monoclonal antibody in treatment of gastrointestinal tumors, *Lancet,* i, 762, 1982.
50. **Ross, A. H., Dietzschold, B., Jackson, D. M., Earley, J. J., Ghrist, B. D. F., Atkinson, B., and Koprowski, H.,** Isolation and amino terminal sequencing of a novel melanoma-associated antigen, *Arch. Biochem. Biophys.,* 242, 540, 1985.

51. **Magnani, J. L., Nilsson, B., Brockhaus, M., Zopf, D., Steplewski, Z., Koprowski, H., and Ginsburg, V.,** A monoclonal antibody-defined antigen associated with gastrointestinal cancer is a ganglioside containing sialylated lacto-N-fucopentaose II, *J. Biol. Chem.,* 257, 14365, 1982.

52. **Thurin, J., Herlyn, M., Hindsgaul, O., Strömberg, N., Karlsson, K.-A., Elder, D., Steplewski, Z., and Koprowski, H.,** Proton NMR and fast-atom bombardment mass spectrometry analysis of the melanoma-associated ganglioside 9-0-acetyl-GD_3, *J. Biol. Chem.,* 260, 14556, 1985.

53. **Kelsoe, G., Reth, M., and Rajewski, K.,** Control of idiotype expression by monoclonal anti-idiotope antibodies, *Immunol. Rev.,* 52, 75, 1980.

54. **Muller, C. A. and Rajewski, K.,** Idiotope regulation by isotype switch variants of two monoclonal antiidiotope antibodies, *J. Exp. Med.,* 159, 758, 1984.

55. **Teitelbaum, D., Rauch, J., Stoller, B. D., and Schwartz, R. S.,** In vivo effects of antibodies against a high frequency idiotype of anti-DNA antibodies in MRL mice, *J. Immunol.,* 132, 1282, 1984.

56. **Urbain, J. and Wuilmart, C.,** Idiotypic regulation in immune networks, *Contemp. Top. Mol. Immunol.,* 8, 113, 1981.

57. **Bona, C. A.,** Parallel sets and the internal image of antigen within the idiotypic network, *Fed. Proc. Fed. Am. Soc. Exp. Biol.,* 43, 2558, 1984.

58. **Stein, K. and Söderström, J.,** Neonatal administration of idiotype or antiidiotype primes for protection against Escherichia coli K13 infection in mice, *J. Exp. Med.,* 160, 1001, 1984.

59. **Ratcliffe, M.,** personal communication.

Chapter 9

ANTI-IDIOTYPE IN THE THERAPY OF B-CELL MALIGNANCIES

M. Raffeld and J. Cossman

TABLE OF CONTENTS

I. INTRODUCTION

Therapeutic intervention with anti-idiotype (anti-Id) antibody offers an attractive biological approach for the clinical management of human B-cell neoplasia. Because idiotype (Id) is essentially a tumor-specific antigen, it theoretically could serve as a target for passive antibody immunotherapy. Furthermore, despite significant advances in the conventional cytotoxic treatment of lymphoma and leukemia, certain types of B-cell neoplasms have been remarkably resistant to eradication by combination chemotherapy.[1] For these neoplasms, the "low-grade" B-cell lymphomas, and leukemias, alternative measures must be instituted to overcome their persistence and typical clinical pattern of chronic recurrence. Since the malignant cells of most low-grade B-cell neoplasms are clonally expanded mature B cells which express surface membrane immunoglobulin (reviewed in Reference 2), they are well suited as candidates for treatment with anti-Id antibody and will be the major focus of this report.

II. HUMAN B-CELL NEOPLASIA — RATIONALE FOR ANTI-ID THERAPY

A. Clinical Characteristics

Human B-cell neoplasms encompass a broad range of clinically and pathologically diverse lymphoproliferative processes.[2] Among the types of B-cell neoplasms, the "low-grade" lymphomas have received the most attention as targets for anti-Id therapy. Their selection for therapeutic trials can be ascribed to several important features: (1) they usually express surface immunoglobulin;[3] (2) they secrete little immunoglobulin into the circulation;[4-6] (3) most are clonal;[2,7,8] (4) in the majority, chemotherapy is not curative and patients die with disease;[1] (5) they are frequently occurring non-Hodgkin's lymphomas in man; and (6) their slow proliferation and protracted clinical course allow time to prepare anti-Id antibody.[1]

By contrast, the precursor B-cell acute lymphoblastic leukemias — common and pre-B types — are immature cells and do not express surface immunoglobulins[2] and, therefore, provide no exposed site for anti-Id recognition. Moreover, the acute lymphoblastic leukemias, large cell lymphomas, and Burkitt's lymphoma are all characterized by rapid growth and can quickly kill the patient without prompt therapeutic measures.[9,10] Conventional treatment of large cell and Burkitt's lymphoma often results in complete sustained remission and, apparently, cure of these "high-grade" neoplasms.[9] Thus, their growth rates and responsiveness to chemotherapy render them generally unavailable to anti-Id treatment.

In plasma cell and plasmacytoid neoplasms, myeloma, and Waldenström's macroglobulinemia, large quantities of immunoglobulin secreted by the neoplastic cells might serve as a barrier to administered anti-Id and block antibody from reaching the cell surface. For these reasons, therapeutic trials have focused on the low-grade, B-cell neoplasms.

B. Clonality — Intratumor Homogeneity

In most instances, B-cell neoplasms are clonal expansions derived from a single parental cell and express a homogeneous immunoglobulin isotype.[3,8,11] Consequently, all malignant cells within the tumor would be expected to share a common Id. As a corollary, neoplasms would each express their own unique idiotype, thereby satisfying one of the requirements for the production of a separate "custom-made" antibody for each patients's lymphoma or leukemia. This unique quality of each B-cell neoplasm is graphically illustrated by Southern blot analysis of restricted lymphoma DNA hybridized with immunoglobulin gene probes to detect variable region gene rearrangements.[8,12] Immunoglobulin gene restriction fragment sizes vary from lymphoma to lymphoma, which presumably reflects utilization of different variable region genes by each neoplasm. The recombinatorial mechanism would generate unique Id for each tumor, recapitulating the process of normal B cells.[13]

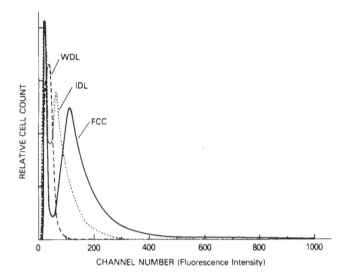

FIGURE 1. Surface immunoglobulin of human low-grade B-cell neoplasms. The three major classes of low-grade B-cell neoplasms — well-differentiated lymphocytic lymphoma (WDL), intermediately differentiated lymphocytic lymphoma (IDL), and follicular center cell lymphomas (FCC) — each express monoclonal surface immunoglobulin. Density of surface immunoglobulin is homogeneously distributed throughout the cell population, as seen by the restricted distribution of fluorescence intensity in these histograms. In addition, the amount of surface immunoglobulin per cell is generally highest in FCC and lowest in WDL.[3]

The picture of monoclonality of human B-cell neoplasms is not so simple, since recent evidence indicates that neoplastic B-cell clones may be genetically unstable.[14-16] The propensity for clonal evolution represents a potential foil for anti-Id treatment and will be addressed in detail in Section V.

C. Immunoglobulin Gene Expression

1. Surface Immunoglobulin

In general, low-grade B-cell lymphoma cells express surface membrane-bound immunoglobulin, usually of IgM isotype, but secrete only minute amounts of immunoglobulin and thereby appear to represent a B-cell maturational stage.[4-7,11] Immunoglobulin is most often of a single heavy and light chain type, and equivalent amounts are distributed on individual cells to produce a homogeneously restricted density of surface immunoglobulin detectable by flow cytometry (Figure 1).[3] Thus, for purposes of recognition by anti-Id, each neoplastic cell would be at equivalent risk.

2. Secreted Immunoglobulin

Although immunoglobulin is not usually detectable by routine clinical electrophoresis, quantitative analysis with anti-Id has revealed that most low-grade B-cell neoplasms secrete at least small amounts of immunoglobulin in serum and urine.[6,17,18] Secreted immunoglobulin detected by anti-Id immunoassays ranges from <0.01 to >500 μg/mℓ in serum[17] and >5 mg/day of light chain in the urine in 40 to 50% of patients.[18] In those cases where secreted immunoglobulin is substantial, it may become a barrier, preventing anti-Id from reaching target cells. Circulating Id can pose a significant problem, which has been only temporarily abrogated by plasmapheresis.[19] Evidently, residual pools of immunoglobulin can be released into the circulation to replace that removed by even whole volume plasmapheresis.

III. PRODUCTION OF ANTI-ID

Sufficient secreted immunoglobulin is not usually available from the low-grade B-cell neoplasms for conventional protein purification. To by-pass this problem several approaches have been successfully applied for obtaining tumor Id for immunization and screening.

A. Preparation of Id for Immunization
1. Papain Digestion of Surface or Total Immunoglobulin
In this approach, described by the Stevenson group,[20,21] Fab fragments resulting from papain digestion are affinity purified on an immunosorbent column and used for immunization. This technique is costly in terms of cells, since it may require in excess of 10^{10} cells, more than could be expected from even a sizable lymph node biopsy.[19]

2. Concentration of Secreted Immunoglobulin from Serum or Urine
When abundant, monoclonal serum immunoglobulin can be readily fractionated from serum for use as an immunogen.[7,22] However, in these cases, the therapeutic value of the anti-Id produced will likely be diminished by the blocking effect of the circulating immunoglobulin. Urinary light chains can be concentrated and have been used to elicit anti-Id antisera reactive with the neoplastic cells in four patients with chronic lymphocytic leukemia (CLL).[23]

3. Production of Secreting Somatic Cell Heterohybrids
Levy and colleagues initially sought to obtain a source of secreted immunoglobulin by fusing human B-cell lymphoma cells with mouse myeloma cell lines. The human immunoglobulin secreted by the heterohybrids could be recovered by affinity column chromatography and then injected into mice to raise murine monoclonal anti-Id.[24-26] Later, secreting human-human hybridomas were similarly prepared.[27] The advantage of this technique is that it provides an unlimited supply of Id for immunization and characterization. However, the approach is cumbersome and time consuming, and hybrids may either be unstable or poorly secretory.[27] The heterohybridoma technique has apparently been replaced by the use of whole cell injections in Levy's laboratory.[19]

4. In Vitro Secretion of Immunoglobulin
As a consequence of results of previous studies concerning immunoglobulin gene expression, we have obtained secreted immunoglobulin from short-term cultures of low-grade B-cell lymphoma cells.[28,29] Although we have found only occasional evidence of spontaneous secretion, secretion is nearly always inducible by the addition of the phorbol ester, tetradecanoyl phorbol acetate (TPA), to CLL cells (Figure 2)[28] or TPA and allogeneic T cells to follicular lymphoma cells (Figure 3).[29] Secretion of immunoglobulin can be ascribed to plasmacytoid differentiation of the cultured cells, since they can be shown to undergo a transition from the transcription of predominantly the 2.4-kb membrane form of μ-chain mRNA to the 2.0-kb mRNA encoding secretory μ-chain, as occurs during the normal differentiation of B cells into plasma cells (Figure 4).[28]

Sufficient secreted immunoglobulin can be collected from 5- to 7-day cultures by affinity chromatography using Sepharose®-coupled antihuman IgM.[14] Recovery is nearly 100%, and the yield ranged from 100 to 1000 μg, which is adequate for immunization and screening (Figure 5). The advantage of this method is that purified Id can be prepared within 1 week of biopsy. However, some cases require screening for appropriate culture conditions, and sufficient cells may be unavailable for final Id production.

5. Injection of Whole Cells or Plasma Membranes
By injecting whole tumor cells, Levy's group has circumvented their prior requirement for secreting heterohybridomas.[27] Anti-Id have been reported to have been prepared for a series of eight patients by this method.[19] The obvious advantage is that Id need not be

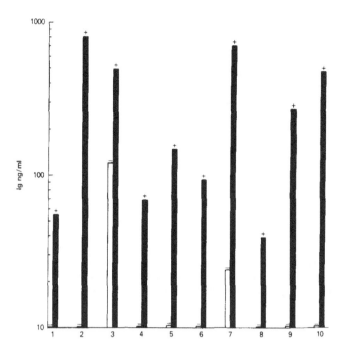

FIGURE 2. Immunoglobulin secretion in vitro by CLL cells. CLL cells from most patients do not spontaneously secrete significant amounts of immunoglobulin in vitro (−). However, cells can be consistently induced to secrete monoclonal immunoglobulin when stimulated with the phorbol ester TPA (+ , solid bars).[28]

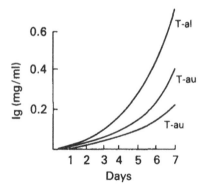

FIGURE 3. Immunoglobulin secretion in vitro in follicular lymphoma. Although little immunoglobulin is spontaneously secreted by follicular B-cell lymphoma cells in vitro, secretion can be induced by the incubation of neoplastic B cells with the naturally accompanying T cells from lymph node cell suspensions in the presence of TPA (T-au). Secretion can be augmented by the depletion of autologous T cells and coculture of neoplastic B cells with allogeneic T cells in the presence of TPA (T-al).[29]

purified, and only 10⁶ cells are required per injection.[27] Alternatively, monoclonal anti-Id against B-cell lymphomas have been elicited by immunization with plasma membranes.[30]

B. Preparation of Anti-Id Antibody
1. Polyclonal Antibody
The Stevensons have shown that specific anti-Id antibody can be raised and isolated from

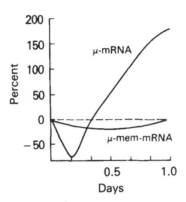

FIGURE 4.   Kinetics of transcription of secretory message of immunoglobulin heavy chain in TPA-stimulated CLL. Short-term exposure of CLL cells to TPA induced a rapid accumulation of cytoplasmic mRNA encoding the secretory form of the immunoglobulin μ chain. This was evident in Northern blot hybridization as well as in the above graphic representation of densitometry of a dot-blot audioradiogram. Despite the rapid rise of total μ-chain mRNA, the RNA hybridized with a μ-membrane probe showed no increase in quantity of message. Thus, CLL cells were fully capable of plasmacytoid differentiation at the level of transformation from a predominant membrane message to that of a secretory message for IgM.[28]

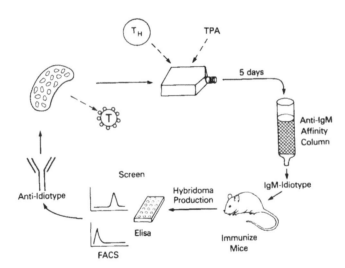

FIGURE 5.   Preparation of anti-Id antibody. Short-term cultures of cell suspensions of lymph nodes bearing low-grade B-cell neoplasms can be induced to secrete immunoglobulin. The secreted immunoglobulin can be purified by column chromatography and used to prepare monoclonal anti-Id antibodies.

the sera of immunized sheep by means of extensive absorption.[31] Polyclonal antisera carry certain advantages over monoclonal anti-Id, since they are more cheaply and rapidly produced and their inherent heterogeneity theoretically provides a spectrum of isotypes and cytotoxicities. Furthermore, ployclonal antisera are likely to contain multiple specificities reactive with different epitopes. The significance of this latter point is underscored by the high mutational rate of Id in B-cell neoplasms and will be addressed in Section V.

2. Monoclonal Antibody

Monoclonal anti-Id have been raised against human B-cell neoplasms using purified Id,

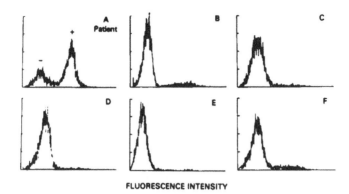

FIGURE 6. Screening for specificity of monoclonal anti-Id. IgM affinity column chromatography was used to isolate idiotypic secretory IgM from culture supernatance from a patient with follicular lymphoma. Secreting hybridomas were screened for specific anti-Id antibody by flow cytometry in which supernatates were tested against the patient lymphoma (A) and five other human B-cell neoplasms of the same immunoglobulin heavy and light chain isotype (B to F). Positive reactivity was detected only against the patient's lymphoma.

plasma membranes, or whole cells as immunogen.[14,26,27,30] Interestingly, idiotopes appear to be selectively immunogenic, since we and others have found that 25 to 30% of hybridomas secreting antiimmunoglobulin are anti-Id (Figure 6).[14,27] This high proportion is attained regardless of immunization with whole cells or purified Id. However, whole cell immunizations do elicit a significantly lower frequency of hybridomas secreting anti-Id (1% with whole cells vs. 15% with purified Id) when techniques were compared within one laboratory.[27] Despite this, the fact that monoclonal anti-Id can be raised by whole cell injections streamlines production by circumventing the isolation and purification of Id.

IV. IN VIVO MONITORING AND TREATMENT

A. Animal Models

When administered to animals inoculated with B-cell neoplasms, anti-Id antibodies have exerted measurable effects. In three systems tested — the guinea pig L_2C leukemia, murine CH1 lymphoma, and BCL_1 leukemia — anti-Id treatment has resulted in reduction of tumor growth and prolonged survival.[32-36] Although antibody therapy was not curative, the animal models are not directly comparable to human low-grade B-cell lymphomas. L_2C, CH1, and BCL_1 are high-grade, rapidly proliferating neoplasms capable of killing the host in a matter of days. Thus, any effect achieved in their growth control is exciting, but might not be applicable to the modest cell kinetics of human low-grade B-cell neoplasms.

B. Treatment of Human Neoplasms

1. Chronic Lymphocytic Leukemia

Using their polyclonal anti-Id, the Stevenson group has reported results of treatment of four patients with CLL.[17,31] Transient reduction of peripheral lymphocyte counts was achieved in three patients. Refractoriness of the leukemias was thought to be due to antigenic modulation as well as the blocking effect of up to 400 μg/mℓ of circulating free idiotypic immunoglobulin.[37] Despite attempts at reducing the level of serum Id by plasmapheresis, no long-term control of disease was attained.[17,37]

2. B-Cell Lymphoma

Two groups have reported treatment results of 13 patients with B-cell lymphoma.[38,39] Unfortunately, only the first patient reported by the Stanford group enjoyed a sustained complete remission.[38] This patient has since remained disease free over the ensuing 42 months.[19] Of the other 12 patients, one had a dramatic decrease in the bulk of tumor which was sustained for at least 3 months; transient but measurable clinical responses were observed in six patients, while five had no response to therapy.[38,39]

Each case represents a unique biological experiment and it is difficult to derive meaningful generalities from these results. The interested reader is urged to study the particular circumstances of each patient in the original descriptions. Toxicities of treatment consisted of fever, chills, dyspnea, rash, headache, nausea, diarrhea, thrombocytopenia, neutropenia, and, in one case, facial palsy.[17,19,39] Toxicities occurred in those patients with circulating tumor cells and free Id or in those who developed an antimouse immunoglobulin immune response. Thus, the toxicities observed seemed to be related to the formation of immune complexes. There appeared to be no correlation between treatment response and the type of B-cell neoplasm, extent of disease involvement, or presence or absence of "B" symptoms.[17,19,39] In addition, the immunoglobulin subclass of mouse anti-Id antibody, the peak serum anti-Id level, the total dose of anti-Id, and the duration of therapy seemed to have no influence on the treatment outcome.[19] One significant correlation was an association between the emergence of an antimouse immunoglobulin response and the subsequent failure of previously responding tumors to continue to regress.[19]

V. IN VIVO MECHANISMS OF ANTI-ID ACTION AND TUMOR ESCAPE

The results of treatment trials strongly suggest that, in some cases, passive anti-Id administration can lower the tumor burden. However, the mechanism by which this might occur has not been elucidated. A number of possibilities have been suggested, including complement-mediated tumor lysis, antibody-dependent cellular cytotoxicity (ADCC), and cellular immunoregulation.

A. Antitumor Activities
1. Complement
In vitro studies have shown that complement mediates lysis of polyclonal anti-Id treated guinea pig L_2C B-lymphoma cells,[40] mouse CH1 cells,[34] and also human CLL cells.[41] However, both the Stanford group and Rankin et al. reported that none of their monoclonal anti-Id antibodies were capable of mediating cell killing in the presence of human complement, even when the IgG subclass was one capable of fixing complement.[38,39] In addition, neither group reported a measurable decrease in serum complement levels following therapy. Finally, Rankin et al. were unable to demonstrate the presence of complement components on the surface of lymphoma cells following administration of anti-Id antibody.[39] Thus, there is no consistent data concerning the role complement plays in tumor lysis in patients treated with anti-Id antibody.

2. Antibody-Dependent Cellular Cytotoxicity
ADCC has been shown to mediate killing of L_2C guinea pig leukemia cells treated with polyclonal anti-Id in vitro.[40] However, there is no evidence to substantiate ADCC as a mechanism of tumor lysis in humans treated with anti-Id.

3. Immunoregulation
The kinetics of response in the only patient who attained a complete remission suggest that the effect of monoclonal anti-Id therapy may not be mediated by direct toxicity, but attributable to immunoregulation. This patient's lymphoma gradually regressed long after anti-Id antibody was detectable in the serum.[19,38] This observation supports the concept that anti-Id might have triggered a sequence of events leading to a delayed immune response.

This intriguing possibility has not yet been formally analyzed and remains a speculative, but important, avenue of investigation.

B. Limitations of Antibody Action

Partial responses indicate that passively administered anti-Id antibodies effectively eliminate some finite fraction of the tumor. Despite not knowing the precise factors involved in tumor lysis, we can still explore mechanisms by which antibody might be prevented from reaching cell targets and mediating individual tumor cell lysis.

Here we discuss factors which might prevent anti-Id antibody from reaching tumor cells, and then we examine features of the tumor cells which might render them invisible to the antibody.

1. Blocking by Circulating Id

An initial obstacle faced by injected anti-Id is the presence of free circulating Id. Although one important criterion for treatment is the lack of clinically detectable paraprotein, many patients with low-grade lymphoma/leukemia have low levels of circulating paraprotein when examined by more sensitive research methods.[6,17-19] In addition, some lymphomas are able to secrete immunoglobulin under the appropriate stimulation.[28,29] In the Stanford series, six patients had detectable serum Id ranging from 2 to 400 μg/mℓ. In patients with small amounts of Id (2 to 15 μg/mℓ), small doses of anti-Id were capable of clearing the Id from the blood.[19] In the two patients with high levels of circulating Id (290 and 400 μg/mℓ), a combination of plasmapheresis and anti-Id was attempted. However, even with plasmapheresis, large doses of anti-Id were required to drive the free Id to undetectable levels. In comparison to patients with no or low free Id levels, the peak serum anti-Id level attained was attenuated, and the half-life of the administered antibody was appreciably shortened. Tumor responses were not observed unless circulating Id was reduced and until measurable levels of circulating administered antibody could be maintained.[19]

2. Penetration

Once anti-Id antibody has overcome circulating free Id, it is then available to reach the tumor cell population throughout the body. Significant blood levels of anti-Id antibody (up to 300 μg/mℓ) have been attained, and antibody could be found in pleural fluid, but did not appear to cross the blood-brain barrier.[19] Antibody readily homes to circulating tumor cells, and clinical studies have demonstrated antibody binding to tumor cells in the blood.[19,37] However, it is less evident that antibody effectively binds tumor cells in other body fluids and solid tissue. To test this, lymph nodes from patients treated by the Stanford group were analyzed for the presence of mouse immunoglobulin following monoclonal antibody treatment. In most patients, mouse antibodies were detected on the tumor cells.[19,42] In addition, Rankin et al. showed morphologic evidence of killing occurring in one patient's lymph node, but did not directly demonstrate the presence of the antibody.[39] Thus, although the data are limited, it seems that antibodies do penetrate the common sites of involvement of low-grade lymphoma/leukemia.

3. Immune Response to Administered Anti-Id

The administered anti-Id antibodies are xenogeneic and, as such, one might predict the development of a host immune response to them. The Stevensons report no such immune responses in their series of four patients treated with sheep polyclonal antibodies.[17] They suggest that disease-associated immunosuppression may be responsible for the lack of a host response.[17] In the two patients treated by Rankin et al., monoclonal antibodies elicited no antimouse responses up to 1 month following treatment.[39] However, in the Stanford series, 5 of 11 patients treated with monoclonal anti-Id elaborated a host response evidenced by the appearance of antimouse immunoglobulin 10 to 24 days after initiation of therapy. The appearance of antimouse antibody was associated with a significant reduction in the half-life of subsequently administered anti-Id, a diminution of peak serum levels of the anti-Id, and lack of additional clinical responses.[19]

4. Modulation

Treatment with anti-Id potentially induces modulation of cell surface immunoglobulin, rendering cell targets less susceptible to immune killing.[17,37] Modulation clearly occurs in the L_2C model and reduces subsequent complement-mediated cell lysis in vitro[43] and may occur in humans in vivo.[37]

C. Tumor Escape
1. Biclonality

Two questions arise regarding the role of clonality in the tumor response to the anti-Id. First, are B-cell neoplasms uniformly clonal and, second, is monoclonality required for effective anti-Id treatment? The answer to the latter question is not fully understood, but is testable as a result of findings of recent studies which have addressed the clonal fidelity and variability of B-cell neoplasms.

In a first approach to this problem, restriction fragments of rearranged immunoglobulin genes were mapped by Southern analysis, and fragment sizes were compared among multiple biopsies from individual patients. In the study reported by Sklar et al.[44] and corroborated by our own series,[47] rearranged immunoglobulin gene restriction fragment patterns were often found to vary from specimen to specimen in individual patients. Variation was observed simultaneously in separate sites sampled, either concurrently or over time, even in biopsies obtained as long as 10 years apart (References 44 and 47). Evidence to date suggests that in most cases the variants represent subclones of the original parental line resulting from secondary DNA deletions or heavy chain switching of previously rearranged genes rather than the emergence of second neoplasms.[45] By contrast, earlier data implied that in some patients secondary, "biclonal" lymphomas might arise which harbor distinct immunoglobulin heavy and light chain gene restriction fragment patterns.[44] However, by itself, genomic blot analysis of rearranged immunoglobulin genes cannot provide absolute evidence of genetic unrelatedness and biclonality, since secondary genetic alterations can occur within a clone to create more restriction fragments.

To confirm this putative lack of genetic relationship between so-called biclonal lymphomas will necessitate further restriction mapping and nucleotide sequencing of the variable region genes. Our own data suggest that true biclonality must be a rare event, since we found no examples within a series of multiple tumor specimens from more than 25 low-grade lymphoma patients.

Of course, secondary reorganization of immunoglobulin genes might not affect the expressed Id if they should occur in the nonproductive allele or outside of idiotope-encoding regions. However, if Id-encoding sequences were altered, loss of anti-Id recognition would be expected.

2. Somatic Mutation

Studies by Raffeld et al.[14] and Meeker et al.[15] have shown the emergence of subpopulations of cells in B-cell lymphomas which no longer reacted with the anti-Id (Figure 7). In contrast to the "biclonal" patients, these two cell populations were shown to be genetically related by restriction fragment analysis (Figure 8).[14,15] It was suggested that somatic mutation of the Id had occurred in these lymphomas, rendering the cell invisible to the anti-Id. Sequence analysis identified multiple somatic mutations in the immunoglobulin gene heavy chain locus, particularly affecting the complementarity-determining region (CDR) II.[16] None of

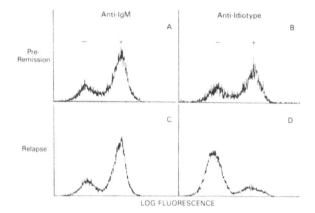

FIGURE 7.   Loss of reactivity with anti-Id antibody in a patient with follicular lymphoma. More than 65% of cells from an early biopsy of follicular lymphoma stained positively with both anti-IgM (A) and anti-Id (B). However, the patient experienced a spontaneous regression of tumor followed by a relapse. At relapse, again the neoplastic B cells were detected by anti-IgM (C), but only a small fraction (22%) of cells could be detected with anti-Id (D).[14]

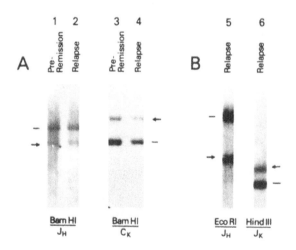

FIGURE 8.   Southern blot analysis of lymphoma specimens before and after spontaneous regression. To determine the genetic relatedness of the two lymphoma biopsies with differential reactivities with anti-Id, DNA was subjected to restriction fragment analysis and probed for immunoglobulin gene rearrangements. Both biopsies show the identical immunoglobulin heavy and capitalite change gene rearrangements. The monoclonal nature of the relapse specimen was confirmed by digestion with second enzymes (B). The demonstration of a single parent cell in both populations suggests that the mechanism for escape from detection by anti-Id was likely due to somatic mutation rather than biclonality.[14]

eight immunoglobulin genes derived from separate subclones of the same tumor were exactly alike. This high mutation rate is disturbing and suggests that subpopulations of tumor cells with altered Id are frequent and might not be susceptible to anti-Id treatment. The frequent somatic mutation rate is not unexpected in these mature B cells, since mutation appears to be a physiological mechanism associated with the generation of antibody diversity during the maturing humoral immune response.[13]

3. Unrearranged Stem Cells

Finally, there is evidence in chronic myelogenous leukemia for the existence of stem cells with unrearranged immunoglobulin genes capable of undergoing multiple rearrangement patterns during lymphoid blast crisis.[46] If an analogous cell exists in low-grade lymphomas, it might provide a reservoir of multiple Id. However, to date no evidence of a stem cell phenomenon has been demonstrated among low-grade B-cell neoplasms.

VI. CONCLUSIONS

Anti-Id therapy potentially faces a large number of problems, ranging from simple penetration to overcoming circulating Id, modulation of surface immunoglobulin, and mosaicism of surface Id. Of these, Id variation is the most difficult to circumvent. One strategy might be to manufacture a cocktail of monoclonal antibodies aimed at covering a range of variant Id. However, one cannot predict variant Id which might arise either spontaneously or during treatment, and it would be impractical to produce sufficient numbers of monoclonal antibodies to keep pace with the natural mutation rate. Alternatively, polyclonal anti-Id could potentially cover many of the variant Id, and their use should not be overlooked.

It is also possible that all Id variants do not have to be covered by the anti-Id treatment if the mechanism of action is mediated by a generalized antitumor response. Alternatively, minor residual variant populations persisting after anti-Id may be susceptible to destruction by a normal host response and, therefore, be of little biologic significance.

Future studies are also likely to expand the clinical settings for the use of anti-Id antibodies by including them as adjuvants in either conventional chemotherapeutic regimens or in combination with other novel immunotherapies. We can also look forward to clinical trials which use the anti-Id to direct adjoined cytotoxic compounds, e.g., ricin or radioisotopes) to the malignant cell population.

REFERENCES

1. **Horning, S. J. and Rosenberg, S. A.,** The natural history of initially untreated low-grade non-Hodgkin's lymphoma, *N.Engl. J. Med.,* 311, 1471, 1984.
2. **Lipford, E. H. and Cossman, J.,** Biological diagnosis of B-cell neoplasia, *Cancer Invest.,* 4, 69, 1986.
3. **Cossman, J., Neckers, L. M., Hsu, S.-M., Longo, D., and Jaffe, E. S.,** Low grade lymphomas: expression of developmentally regulated B-cell antigens, *Am. J. Pathol.,* 115, 117, 1984.
4. **Alexanian, R.,** Monoclonal gammopathy in lymphoma, *Arch. Intern. Med.,* 135, 62, 1975.
5. **Axelsson, U., Bachmann, R., and Hallen, J.,** Frequency of pathologic protein (M components) in 6995 sera from an adult population, *Acta Med. Scand.,* 179, 235, 1966.
6. **Stevenson, F. K., Hamblin, T. J., Stevenson, G. T., and Tutt, A. L.,** Extracellular idiotypic immunoglobulin arising from human leukemic B lymphocytes, *J. Exp. Med.,* 152, 1484, 1980.
7. **Schroer, K. R., Briles, D. E., Van Boxel, J. A., and Davie, J. M.,** Idiotypic uniformity of cell surface immunoglobulin in chronic lymphocytic leukemia: evidence for monoclonal proliferation, *J. Exp. Med.,* 140, 1416, 1974.
8. **Arnold, A., Cossman, J., Bakhshi, A., Jaffe, E. S., Waldmann, T. A., and Korsmeyer, S. J.,** Immunoglobulin-gene rearrangements as unique clonal markers in human lymphoid neoplasms, *N.Engl. J. Med.,* 309, 1593, 1983.
9. **DeVita, V. T., Canellos, G. P., Chabner, B. A., Schein, P. A., Hubbard, S. P., and Young, R. C.,** Advanced diffuse histiocytic lymphoma: a potentially curable disease, *Lancet,* 1, 248, 1975.
10. **Fisher, R., Hubbard, S. M., DeVita, V. T., Berard, C. W., Wesley, R., Cossman, J., and Young, R. C.,** Factors predicting long term survival in diffuse-mixed, histiocytic, or undifferentiated lymphoma, *Blood,* 58, 45, 1981.
11. **Preud'homme, J. L. and Seligman, M.,** Surface immunoglobulin on human lymphoid cells, in *Progress in Clinical Immunology,* Schwartz, R. S., Ed., Grune & Stratton, New York, 1974, 121.
12. **Cleary, M. L., Chao, J., Warnke, R., and Sklar, J.,** Immunoglobulin gene rearrangement as a diagnostic criterion of B cell lymphoma, *Proc. Natl. Acad. Sci. U.S.A.,* 81, 593, 1984.
13. **Tonegawa, S.,** Somatic generation of antibody diversity, *Nature,* 302, 575, 1983.
14. **Raffeld, M., Neckers, L., Longo, D., and Cossman, J.,** Spontaneous alteration of idiotype in a monoclonal B-cell lymphoma: escape from detection by anti-idiotype, *N.Engl. J. Med.,* 312, 1655, 1985.

15. **Meeker, T., Lowder, J., Cleary, M. L., Stewart, S., Warnke, R., Sklar, J., and Levy, R.,** Emergence of idiotype variants during treatment of B-cell lymphoma with anti-idiotype antibodies, *N.Engl. J. Med.,* 312, 1658, 1985.

16. **Cleary, M. L., Meeker, T. C., Levy, S., Lee, E., Trela, M., Sklar, J., and Levy, R.,** Clustering of extensive somatic mutations in the variable region of an immunoglobulin heavy chain gene from a human B cell lymphoma, *Cell,* 44, 97, 1986.

17. **Stevenson, G. T. and Stevenson, F. K.,** Treatment of lymphoid tumors with anti-idiotype antibodies, *Springer Semin. Immunopathol.,* 6, 99, 1983.

18. **Pierson, J., Darley, T., Stevenson, G. T., and Virji, M.,** Monoclonal immunoglobulin light chains in urine of patients with lymphoma, *Br. J. Cancer,* 41, 681, 1980.

19. **Meeker, T. C., Lowder, J., Molmeg, D. G., Miller, R. A., Thielemans, K., Warnke, R., and Levy, R.,** A clinical trial of anti-idiotype therapy for B cell malignancy, *Blood,* 65, 1349, 1985.

20. **Stevenson, G. T. and Stevenson, F. K.,** Antibody to a molecularly defined antigen confined to a tumor cell surface, *Nature,* 254, 714, 1975.

21. **Hough, B. W., Eady, R. P., Hamblin, T. J., Stevenson, F. K., and Stevenson, G. T.,** Anti-idiotype sera raised against surface immunoglobulin of human neoplastic lymphocytes, *J. Exp. Med.,* 144, 960, 1976.

22. **Wernet, P., Feizi, T., and Kunkel, H. G.,** Idiotype determinants of immunoglobulin M detected on the surface of human lymphocytes by cytoxicity assays, *J. Exp. Med.,* 136, 650, 1972.

23. **Tutt, A. L., Stevenson, F. K., Smith, J. L., and Stevenson, G. T.,** Antibodies against urinary light chain idiotypes as agents for detection and destruction of human neoplastic B lymphocytes, *J. Immunol.,* 131, 3058, 1983.

24. **Levy, R. and Dilley, J.,** Rescue of immunoglobulin secretion from human neoplastic lymphoid cells by somatic cell hybridization, *Proc. Natl. Acad. Sci. U.S.A.,* 75, 2411, 1978.

25. **Brown, S., Dilley, J., and Levy, R.,** Immunoglobulin secretion by mouse X human hybridomas: an approach for the production of anti-idiotype reagents useful in monitoring patients with B cell lymphoma, *J. Immunol.,* 125, 1037, 1980.

26. **Hatzubai, A., Maloney, D. G., and Levy, R.,** The use of a monoclonal anti-idiotype antibody to study the biology of a human B cell lymphoma, *J. Immunol.,* 126, 2397, 1981.

27. **Thielemans, K., Maloney, D. G., Meeker, T., Fujimoto, J., Doss, C., Warnke, R., Bindl, J., Gralow, J., Miller, R. A., and Levy, R.,** Strategies for production of monoclonal anti-idiotype antibodies against human B-cell lymphomas, *J. Immunol.,* 133, 495, 1984.

28. **Cossman, J., Neckers, L. M., Braziel, R. M., Trepel, J. B., Korsmeyer, S. J., and Bakhshi, A.,** In vitro enhancement of immunoglobulin gene expression in chronic lymphocytic leukemia, *J. Clin. Invest.,* 73, 587, 1984.

29. **Braziel, R. M., Sussman, E., Jaffe, E. S., Neckers, L. M., and Cossman, J.,** Induction of immuno-globulin secretion in follicular non-Hodgkin's lymphomas: role of immunoregulatory T cells, *Blood,* 66, 128, 1985.

30. **Rankin, E. M. and Hekman, A.,** Mouse monoclonal antibodies against the idiotype of human B cell non-Hodgkin's lymphomas: production, characterization and use to monitor the progress of disease, *Eur. J. Immunol.,* 14, 1119, 1984.

31. **Hamblin, T. J., Abdul-Ahad, A. K., Gordon, J., Stevenson, F. K., and Stevenson, G. T.,** Preliminary experience in treating lymphocytic leukaemia with antibody to immunoglobulin on the cell surfaces, *Br. J. Cancer,* 42, 495, 1980.

32. **Glennie, M. J. and Stevenson, G. T.,** Univalent antibodies kill tumor cells in vitro and in vivo, *Nature,* 295, 712, 1982.

33. **Stevenson, G. T., Elliott, E. V., and Stevenson, F. K.,** Idiotypic determinants on the surface immu-noglobulin of neoplastic lymphocytes: a therapeutic target, *Fed. Proc. Fed. Am. Soc. Exp. Biol.,* 36, 2268, 1977.

34. **Haughton, G., Lanier, L. L., Babcock, G. F., and Lynes, M. A.,** Antigen-induced murine B cell lymphomas. II. Exploitation of the surface idiotype as tumor specific antigen, *J. Immunol.,* 121, 2358, 1978.

35. **Strober, S., Gronowicz, E. S., Knapp, M. R., Slavin, S., Vitetta, E. S., Warnke, R. A., Kotzin, B., and Schroder, J.,** Immunobiology of a spontaneous murine B-cell leukemia (BCL₁), *Immunol. Rev.,* 48, 169, 1979.

36. **Krolick, K. A., Uhr, J. W., Slavin, S., and Vitetta, E. S.,** In vivo therapy of a murine B cell tumor (BCL₁) using antibody-ricin A chain immunotoxins, *J. Exp. Med.,* 155, 1797, 1982.

37. **Gordon, J., Abdul-Ahad, A. K., Hamblin, T. J., Stevenson, F. K., and Stevenson, G. T.,** Mechanism of tumour cell escape encountered in treating lymphocytic leukemia with anti-idiotypic antibody, *Br. J. Cancer,* 49, 547, 1984.

38. **Miller, R. A., Maloney, D. G., Warnke, R., and Levy, R.,** Treatment of B-cell lymphoma with monoclonal anti-idiotype antibody, *N.Engl. J. Med.,* 306, 517, 1982.

39. **Rankin, E. M., Hekman, A., Somers, R., and ten Bokkel Huinink, W.,** Treatment of two patients with B cell lymphoma with monoclonal anti-idiotype antibodies, *Blood,* 65, 1373, 1985.
40. **Stevenson, F. K., Elliott, E. V., and Stevenson, G. T.,** Some effects on leukaemic B lymphocytes of antibodies to defined regions of their surface immunoglobulin, *Immunology,* 32, 549, 1977.
41. **Hamblin, T., Gordon, J., Stevenson, F., and Stevenson, G.,** Reduction of blocking factor to immunotherapy by plasma exchange, in *Sonderruck aus Plasma Exchange,* Sieberth, H. G., Ed., Schattauer Verlag, Stuttgart, 1980, 387.
42. **Garcia, C. F., Lowder, J., Meeker, T. C., Bindl, J., Levy, R., and Warnke, R.,** Differences in "host infiltrates" among lymphoma patients treated with anti-idiotype antibodies: correlation with treatment response, *J. Immunol.,* 135, 4252, 1985.
43. **Gordon, J., Robinson, D. S. F., and Stevenson, G. T.,** Antigenic modulation of lymphocytic surface immunoglobulin yielding resistance to complement-mediated lysis. I. Characterization with syngeneic and xenogeneic complements, *Immunology,* 42, 7, 1981.
44. **Sklar, J., Cleary, M. L., Thielemans, K., Gralow, J., Warnke, R., and Levy, R.,** Biclonal B-cell lymphoma, *N.Engl. J. Med.,* 311, 20, 1984.
45. **Siegelman, M. H., Cleary, M. L., Warnke, R., and Sklar, J.,** Frequent biclonality and Ig gene alterations among B cell lymphomas that show multiple histological forms, *J. Exp. Med.,* 161, 850, 1985.
46. **Bakhshi, A., Minowada, J., Arnold, A., Cossman, J., Jensen, J. P., Whang-Peng, J., Waldmann, T. A., and Korsmeyer, S. J.,** Lymphoid blast crises of chronic myelogenous leukemia represent stages in the development of B-cell precursors, *N.Engl. J. Med.,* 309, 826, 1983.
47. **Raffeld, M. and Cossman, J.,** unpublished data.

Chapter 10

THE BIOLOGICAL SIGNIFICANCE OF ANTI-IDIOTYPIC AUTOIMMUNE REACTIONS TO HLA

N. Suciu-Foca and D. W. King

TABLE OF CONTENTS

I. INTRODUCTION

Implicit in the notion of immune recognition is the concept that T and B lymphocytes are able to discriminate between self- and nonself (or altered-self)-antigens. Recognition of self is largely based on the identification of antigens encoded by genes at the major histocompatibility complex (MHC).[1-3] The MHC is a chromosomal region that comprises a number of closely linked and highly polymorphic genes which are involved in normal immune response and cellular recognition function. The products of these genes (H2 antigens in mice and HLA in humans) play a major role in T-cell recognition of nonself-antigens. Thus, cytolytic T cells recognize cell membranes altered by viruses, chemical substances, or mutations, only in the context of self-Class I HLA molecules.[2,3] Regulatory T cells (helper, suppressor, or amplifiers) are thought to recognize at the surface of antigen-presenting cells a complex consisting of nominal antigen associated with Class II MHC determinants.[1-3] T cells or T-cell-dependent/B-cell responses are initiated only following recognition of self-MHC plus X complexes. The phenomenon of MHC-restricted immune recognition does not apply, however, to allogeneic MHC antigens. A large proportion of T cells, in most mammalian species, displays the ability of recognizing specifically any allelic variant of self-MHC.

Alloreactivity is expressed in an individual's ability to reject a graft from another member of the same species. This is, in fact, the earliest immune response that can be demonstrated both in philogeny (as it is already present in fish) and in ontogeny (since it is displayed by embryonal thymocytes).[3] The biological significance of alloreactivity is still unknown.

Numerous studies suggest that alloreactivity reflects the high degree of cross reactivity between allelic variants of the MHC and autologous MHC antigens complexed with foreign antigens. If so, the repertoire of alloreactive cells includes lymphocytes endowed with the capacity of recognizing foreign antigens in the context of self-MHC antigens. The structural basis of such cross reactivity is not well understood, yet it is presumed that T cells utilize the same antigen receptor for allogeneic MHC and for recognition of self plus X complexes.[4-6]

We have suggested in previous work that selective pressures generating the extraordinary polymorphism of MHC and of genes encoding anti-MHC immune responses also resulted in the development of mechanisms which suppress rejection of the fetus in outbred animals.

We have further postulated that such mechanisms may consist of network regulation of alloimmune responses and that their existence is a necessity for the survival of the species.

II. IMMUNE RESPONSES ARE SUBJECTED TO NETWORK REGULATION

Certain postulates of the network hypothesis are accepted almost universally. Antigenic determinants unique to a small set of antibody molecules have been referred to as idiotypes (Id) or idiotypic determinants which are recognized by a complementary set of anti-idiotypic (anti-Id) antibodies. Id define the variable region of the immunoglobulin molecule and may serve as penotypic markers for the V region. The antigen-binding region of the antibody molecule has been referred to as a paratope. The complementarity of structural conformations on paratopes and corresponding epitopes plays a role in antigen binding to the antibody molecules and antigen mimicry by anti-antibodies (anti-Id antibodies or Ab$_2$).[7-16]

The immune response to most antigens involves several clones of reactive T and B lymphocytes, expressing many idiotypic specificities or antigenic differences in the V region. After antigenic stimulation, however, only certain clones expressing a prevailing or dominant Id may proliferate. Idiotypic cross reactions are not uncommon in both inbred and outbred animals, probably as a result of preservation and sharing of certain germline genes.[7-16]

The network theory predicts that autoimmune responses to self-Id form the basis of immunoregulatory network systems in which homeostasis is preserved through a functional assembly of Id-anti-Id interactions. Jerne suggested that the regulation of this system accounts for the various modes of the immune response, i.e., steady (preimmune) state, immune state (when expansion of clones with fitting paratopes takes place), and postimmune state (when anti-idiotypic suppression of the immune response and/or immune memory develop).[7,8]

Since each animal can make Id antibodies against the virtually unlimited antigenic universe, and since each Id can induce anti-Id antibodies, Id and complementary anti-Id should coexist within the individual's immune repertoire. The immune system could, thus, contain imperfect internal images of the external universe which may substitute for antigen under certain conditions. This extreme, yet provocative, view implies that nonself-antigens are not perceived by the immune system as "foreign" or new, since the system recognizes these structures continuously in its complementary circuits. The antigen may only perturb the system until it reaches a new steady state. Immunoregulatory circuits based on idiotypic interactions between and within the T- and B-cell compartment may then determine the type, duration, and magnitude of any immune state.[8-13]

In the following we will review our data, suggesting that anti-idiotypic immune responses to nonself-MHC are involved in maternal tolerance to the fetus and in allograft acceptance. We will further argue that suppression of the response is a necessity for the survival of the species and that such mechanisms have evolved from anti-idiotypic reactions directed primarily against clones reactive to self-MHC.

III. ANTI-IDIOTYPIC IMMUNE RESPONSES TO HLA IN PREGNANCY

A. Development of Anti-HLA Anti-Id during Pregnancy

In mammalian reproduction, which entails natural transplantation and maintenance of the conceptus as a successful allograft, histocompatibility genes have been shown to play a significant and unexpected role: incompatibility favors both its chances of implantation and its subsequent growth.[17]

It has long been recognized that the fetoplacental unit is a highly successful allograft, yet the mechanism which prevents its rejection, in spite of obvious sensitization of the mother to the MHC antigens of the fetus, is still unclear. It is quite possible that active immunoregulatory responses mediated by "blocking" antibodies, by antigen/antibody complexes, or by suppressor T cells are involved.[17]

We have suggested in previous studies that the basic mechanism which secures maternal tolerance of the semiallogeneic concept, and by implication the preservation of MHC polymorphism in tandem with T-cell alloreactivity, resides in the development of anti-idiotypic autoimmunity to MHC.[18-20] Our concept is supported by several findings.

HLA antigens are highly immunogenic and induce the formation of anti-HLA antibodies. Thus, in pregnant women anti-HLA antibodies are usually detectable during the third trimester of pregnancy and can be shown to persist for variable periods of time which range from 1 month to several years following delivery.[1]

In our earliest studies, we showed that the F(ab')$_2$ fraction of sera obtained at a time when HLA antibodies had disappeared from the circulation was capable of blocking the cytotoxic activity of antibodies present in earlier bleeding from the same woman.[21,22] The blocking serum, containing the putative anti-Id (Ab$_2$), also inhibits the cytotoxic activity of certain sera from unrelated individuals which had been immunized to the same HLA antigen. This blocking activity displayed an exquisite specificity, since it rarely expanded beyond the limits predicted on the basis of the public or private epitopes carried by the immunizing haplotype of the child.[22]

The generality of this phenomenon has been confirmed in investigations which have showed that of 43 sera obtained at the time of delivery from primipara, 36 had anti-anti-

Table 1
CYTOFLUOROMETRIC DETERMINATIONS OF INHIBITION OF Ab₁ BY AUTOLOGOUS Ab₂ FROM SERA OBTAINED DURING PREGNANCY

Titer of Ab₁ (placental serum)	No Ab₂ added (%)	Paternal T lymphoblasts reacting with placental serum in the presence of F(ab')₂ obtained from sera collected prior to or following delivery (%)			
		+ Ab₂ (− 5 months)	+ Ab₂ (− 2 months)	+ Ab₂ (Delivery)	+ Ab₂ (5 months)
1:2	91	1.1	1.4	88	74
1:4	84	0.4	1.4	76	68
1:8	93	1.2	1.3	68	70
1:16	86	1.4	3.1	65	52
1:32	81	0.8	2.7	40	31
1:64	63	1.5	3.6	36	14

HLA antibodies (Ab_2) which specifically inhibit the anti-HLA activity of sera reacting with their husbands' Class I or II antigens. Nine of the Ab_2-positive sera also contained anti-HLA (Ab_1) activity, demonstrating that the two generations of antibodies coexist within the immunoglobulin pool.

This suggests that responsiveness or nonresponsiveness to HLA merely reflects the prevalence of Ab_1 or Ab_2. It is apparent that if Ab_2 plays a biological role, securing maternal tolerance to the fetus, it should prevail early during pregnancy. We have confirmed this possibility in a study in which we investigated the blocking activity of sequential sera obtained from a woman during gestation. Table 1 illustrates such a study on a primiparous woman. This woman developed anti-HLA-A2 (1:64 titer) and anti-HLA-DR2 (1:8 titer) Ab_1 at the time of delivery.[23,24] Sera collected 4 and 2 months earlier, i.e., during the 5th and 7th months of pregnancy, completely blocked the AB_1 found in the undiluted placental serum. The placental serum (depleted of Ab_1) self-inhibited, however, its nonabsorbed counterpart by only 50% even at high (1:16 to 1:64) dilutions. Following delivery, when the titer of Ab_1 declined, there was a second rise in the titer of Ab_2. This suggests that the anti-HLA anti-Id reinstitute the steady state by suppressing Ab_1 formation.

B. Cross-Reactive Id in Conventional Alloantisera

The serologic response to an incompatible MHC haplotype involves the production of antibodies to both Class I and II antigens, as well as to a large array of determinants on each of these molecules. The response to a single epitope is expected to involve multiple clones of reactive cells. Subsequently, the possibility of recognizing anti-Id antibodies to all relevant Id among such heterogeneous populations of antibodies is remote. Furthermore, because of the diversity of alloimmune responses, it is not clear that anti-Id which recognize anti-MHC antibodies in the autologous system would react with a significant percentage of conventional alloantibodies.

Therefore, initial studies were aimed to determine whether anti-Id antibodies recognizing several distinct Class I and II antigens cross reacted with antibodies present in alloantisera. In these studies, the F(ab')₂ fraction obtained from a serum known to inhibit Ab_1 anti-HLA in the syngeneic system was tested for its ability to block the cytotoxic activity of anti-HLA alloantisera exhibiting the same or different specificities.

Blocking of cytotoxicity was ascertained using a well-characterized HLA reference panel of cells. We found that both in the case of Class I and II antigens, anti-idiotypic reagents blocked anti-HLA sera only if the latter were directed against the HLA antibodies which had, presumably, stimulated the production of the respective Ab_2. Although these data indicate that Ab_2 recognizes germline-encoded, cross-reactive Id, the degree of inhibition of various alloantisera and the number of alloantisera that were blocked by different Ab_2 reagents varied widely.[22,25] There was also heterogeneity among target cells with respect to susceptibility to lysis. Thus, when various dilutions of Ab_1 were assayed in the presence of constant concentrations of Ab_2, lysis of some targets was blocked by Ab_2 even when Ab_1 was not diluted, while other targets remained susceptible to lysis over a wide range of Ab_1 dilutions. This suggested heterogeneity in the structure and expression of distinct HLA antigens on cells from different individuals and/or heterogeneity within the population of Ab_1 causing specific lysis.

Since alloantisera from unrelated individuals were inhibited by a given anti-idiotypic reagent, we performed a systematic search for cross-reactive Id. The assumption was made that alloantisera which display very similar patterns of reactivity, i.e., which have a high internal correlation, may share cross-reactive or regulatory Id which could be recognized using a certain anti-idiotypic antiserum. Thus, within the framework of international histocompatibility workshops, we tried to determine whether anti-HLA-DR2 and anti-HLA-DR4 reagents collected from individuals of different ethnic backgrounds contain Id antibodies which may be recognized by anti-Id antibodies (to DR2 and DR4, respectively) that had been characterized in our laboratory. Our anti-Id antibodies blocked some (3 out of 15) of the anti-DR4 and some anti-DR2 (4 out of 16) sera.[25] These sera may share a regulatory Id encoded by germline genes. However, the internal correlation between the workshop antisera which shared the regulatory Id was not higher than the correlation with sera in which this cross-reactive Id could not be detected. This indicated that a whole array of private Id contributes to what may ultimately represent the specificity of an alloantiserum.[25] The answer to this question can best be answered using human monoclonal anti-HLA and anti-anti-HLA antibodies.

C. Autoantibodies Which Recognize Specifically Alloactivated T Lymphoblasts

Earlier studies have shown that F_1 rats immunized with alloreactive T-cell population produced by immunization between parental strains exhibit suppressed responses to alloantigens.[26,27] Attempts to reproduce these studies have failed, however, in many laboratories. The most likely reasons for these failures are the heterogeneity of the population of cells used for immunization as well as the heterogeneity of immune responses induced by the idiotypic receptors for antigen, presumably, expressed by such cells.

With the advent of T-cell cloning and of monoclonal antibodies specific for T-cell antigen receptor of T-cell clones, it has become evident that idiotypic (or clonotypic) receptors for antigen are, indeed, present on alloreactive cells, and that blocking of the receptor by antireceptor antibodies inhibits T-cell function.[28-30]

Progress generated by such studies and particularly by their contribution to understanding the molecular structure of the T-cell receptor for antigen has further shown that, although similar, T- and B-cell receptors are products of different genes. They must, therefore, bear different Id on the variable region of the constitutive chains.[31-41]

The development of anti-T-cell receptor autoantibodies seems to occur, however, spontaneously in nature and plays an important role in maternal tolerance to the fetus. We found that similar to murine monoclonal antibodies which recognize the T-cell antigen receptor on human T-cell clones,[28-30] autoantibodies present in sera from pregnant women react specifically with T-cell clones sensitized to the paternal HLA haplotype of the child.[21] The $F(ab')_2$ fraction of such sera triggers clonal proliferation in the absence of the antigen and blocks the capacity of the reactive clones to kill target lymphocytes from the sensitizing donor. Furthermore, there is comodulation between the molecules detected by the autologus $F(ab')_2$ and anti-T3 antibodies, reinforcing the view that the $F(ab')_2$ contains anti-T-cell-receptor antibodies. Such antibodies seem, therefore, to react with the T-cell receptor for alloantigen and provide a mechanism for down regulation of T-cell alloactivation.[42,56]

In essence, our data suggest that maternal tolerance to the fetus is the result of active recognition and regulation of cellular and humoral immune responses to HLA. The hypothesis that anti-idiotypic immunity plays a role in this phenomenon finds support in the successful outcome of clinical trials in which women with (spontaneous) abortions were treated by immunization with paternal leukocytes. The success rate in terms of preserving pregnancy in habitual aborters that have been transfused with the husband's leukocytes was found to be greater than 70%.[43]

IV. ANTI-ID ANTIBODIES TO HLA IN TRANSFUSED AND TRANSPLANTED PATIENTS

A. Anti-Id Antibodies in Patients with No History of Antibodies to the HLA Antigens of the Donor

Organ allografts can be considered the clinical counterpart of nature's transplantation experiment.

The highest rate of transplant survival results in patients who have received pretransplant transfusions and have then been treated with cyclosporine. The beneficial effect of transfusions is still poorly understood. It has been suggested that transfusions may induce suppressor cells, enhancing antibodies or donor-specific cytotoxic cells which may be deleted upon initiation of the immunosuppressive treatment.

In an attempt to determine whether the beneficial effect of transfusions is related to the development of anti-Id antibodies, we have monitored the development of anti-anti-HLA antibodies in patients who have been preconditioned for transplantation by multiple transfusions with blood from their prospective donor. Ab_2 was tested in recipients who, following three transfusions (administered at 2-week intervals), have not developed anti-HLA antibodies against their respective haploidentical transfusion donors and were subsequently transplanted with this donor's kidney. The specificity of Ab_2, as ascertained by blocking of reference (Ab_1) anti-HLA reagents, and the dynamics of their development in relationship to the transfusion protocol were studied.[44] In most patients, anti-anti-HLA-DR antibodies (Ab_2) occurred 2 weeks following the first transfusion. In a few cases (<20%), Ab_2 was detectable only following the third transfusion of blood from the same donor. Of the patients developing Ab_2 following the first transfusion, 30% showed an increase in titer after the second and third transfusions, 60% showed minor fluctuations, and 10% maintained the same level of Ab_2 activity.[44] Four weeks after the last transfusion, those patients who had not developed cytoxic antibodies were transplanted. These patients showed significant Ab_2 and the (N = 14) renal allografts maintain excellent function for as long as 3 years following transplantation.

B. Anti-Id Antibodies in Patients with a History of Antibodies to the Donor's HLA Antigens

An undesirable effect of the multiple transfusion protocol resides in the induction of anti-HLA antibodies in a relatively large number (>60%) of the patients. For the last 30 years, it was dogmatically accepted that such patients should not receive a graft carrying the HLA antigen against which their antibodies are directed. Instead, a large body of literature had shown that patients with antidonor HLA antibodies will reject the graft hyperacutely. The antibodies attach to the endothelium of capillaries in the renal cortex, the complement system is called into play, and polymorphonuclear leukocytes forming bridge-like structures adhere to the endothelium. When the endothelium is denuded, a rough surface is exposed, allowing platelets to attach, form a thrombus, obstruct the vessels, and cause widespread cortical

Table 2
OUTCOME OF RENAL ALLOGRAFTS IN PATIENTS WITH A HISTORY OF ANTI-DONOR-HLA SENSITIZATION: CORRELATION OF TRANSPLANT SURVIVAL WITH Ab_2 AND OF TRANSPLANT REJECTION WITH Ab_3

Outcome of renal allograft	Total no. patients	Antidonor HLA-antibody at the time of transplantation					
		Ab_1		Ab_2		Ab_3	
		Present	Absent	Present	Absent	Present	Absent
Accelerated rejection	9	0	9	0	9	7	2
Successful	10	0	10	9	1	0	10

necrosis. To avoid hyperacute and acute fulminating rejections, the practice had been accepted of transplanting only patients with no history of sensitization to the donor's HLA antigen.

The shortage of donors, the difficulty of finding HLA-compatible donors for highly immunized patients, and possibly, also, the lack of ready access to the patient's immunologic follow-up have led several groups to transplant individuals with antidonor HLA antibodies. These "accidents" led to the unexpected finding that although some patients rejected the graft violently, other patients who had antidonor antibodies in the past, but not at the time of transplantation, had excellent graft survival. It, therefore, appeared that these patients had no immune memory to the respective HLA alloantigen.[45]

An alternative explanation was that the patient had developed anti-Id autoantibodies which blocked the idiotypic antidonor HLA antibodies. We investigated 19 patients with an historically positive cross-match to their kidney allograft donor. Nine of these recipients rejected the graft, while ten had an uneventful follow-up. Of these ten patients who were successfully transplanted, nine had anti-Id antibodies which blocked an earlier autologous serum with antidonor HLA activity. There was, thus, a significant association between the presence of Ab_2 and tolerance to the graft ($P < 0.001$). None of the nine patients who rejected the graft had anti-Id antibodies (Table 2). Furthermore, the serum obtained at the time of transplantation from seven of these nine patients with hyperacute rejection potentiated rather than blocked the cytotoxic activity of the earlier, historically positive sample. There was, thus, a high correlation between Ab_1-potentiating activity and rejection ($P > 0.001$). The Ab_1-potentiating activity was present in the IgG fraction of the serum and could not be explained by subthreshold levels of anti-HLA antibodies. Antidonor activity was not detected at any concentration of IgG (Table 2).

We postulated that Ab_1-potentiating antibodies might be, in fact, anti-anti-anti-HLA antibodies (or Ab_3) which bind to Ab_2, thus, preventing its reaction with Ab_1.[44] To determine whether Ab_1-potentiating antibodies may also explain early rejections in patients with no preformed antidonor antibodies, we examined sera obtained at the time of transplantation from 12 patients with acute fulminating rejections of renal allografts. Eight of these patients had Ab_1-potentiating antibodies, i.e., their serum augmented the cytotoxic activity of homologous antisera reacting specifically with the donor's lymphocytes. The IgG fraction of these sera did not behave like Ab_1 or internal image of Ab_1, since it did not react with the donor's HLA antigens, although it enhanced specifically the cytotoxic activity of anti-HLA antibodies reacting with donor cells.

Study of a large number of multitransfused patients showed that, frequently, sera with no anti-HLA alloantibodies (Ab_1) contained, in fact, anti-Id antibodies to a certain HLA antigen such as HLA-A1 and Ab_3-type antibodies to another antigen such as HLA-A2, for example.

Table 3
FREQUENCY OF NATURAL AUTO- AND
ALLOANTIBODIES TO HLA IN
CYCLOSPORINE-TREATED RECIPIENTS
OF HEART AND RENAL ALLOGRAFTS

No. of patients	Total no. of sera	No. patients (no. sera) with natural anti-HLA	
		Autoantibodies	Alloantibodies
129	1780	9(13)	36(482)

Histocompatibility testing should, therefore, include not only cross-matching of patient serum with donor lymphocytes so as to preclude the presence of anti-HLA antibodies (Ab_1), but also testing of Ab_2 and Ab_3 antibodies. If anti-antidonor Ab_2 are present at the time of transplantation, the patient could be safely grafted. Patients with Ab_3, however, are at the same risk of accelerated rejection as are those individuals with Ab_1 (antidonor HLA antibodies). "Matching" of alloimmunized individuals for Ab_2 may solve the problem of finding suitable donors for highly sensitized individuals.

V. NATURAL AUTO- AND ALLO-ANTIBODIES TO HLA

As mentioned in the introduction, network concepts led to the view that the immune system comprises imperfect images (or homobodies) of antigens. If the repertoire of paratopes to all possible epitopes is complete, the system should contain imperfect internal images of the external universe of antigens, including MHC. Such internal images could substitute for antigen under certain conditions. In the world of anti-HLA immunity, this implies that Ab_2 can be the internal image of "allo-MHC" (or its distorted image) and Ab_3 can be the image of Ab_1 (as seems to be the case for the Ab_1-potentiating antibodies which we discussed above).

Natural alloantibodies directed against MHC antigens have been found in about 25% of C57B mice, and their appearance has been related to the aging process.[46] Only a few cases of natural anti-HLA alloantibodies have been reported in humans.[47] A recent study showed, however, that almost 1% of normal blood donors have natural anti-HLA antibodies of IgM isotype, and that these antibodies are relatively weak and most often directed against the HLA-B8 antigen.[47] It was suggested that these antibodies are induced through sensitization to environmental agents which cross react with MHC, or by sensitization to altered self-MHC antigens.[47] Although autoantibodies to numerous antigens were detected in human sera, apparently no autoantibodies to HLA have been previously reported.

In a monthly survey of sera obtained from cyclosporine-treated recipients of heart and renal allografts, we found, however, that a significant number of patients sporadically develop anti-HLA alloantibodies of exquisite specificity (Table 3). These antibodies reacted with HLA antigens which were distinct from those of the graft. Furthermore, 7% of the patients developed at one time or another anti-HLA-Class I or anti-HLA-Class II autoantibodies. There was no detectable cross reactivity between the private and public antigens of the graft and the anti-HLA auto- or alloantibodies which the patient developed. These types of natural auto- and alloantibodies to HLA were found to persist in the circulation for relatively short periods of time, rarely exceeding 1 to 2 months (i.e., they were rarely found in more than two subsequent bleedings). They reoccurred sporadically at later times, in some cases, although in other cases the patient developed antibodies of an entirely different specificity.

GENERATION OF ANTI MHC AUTOANTIBODIES

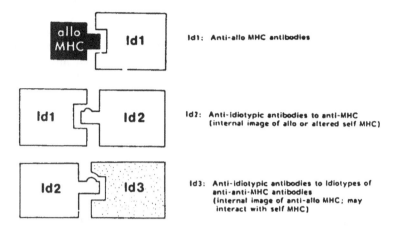

Id1: Anti-allo MHC antibodies

Id2: Anti-idiotypic antibodies to anti-MHC
(internal image of allo or altered self MHC)

Id3: Anti-idiotypic antibodies to idiotypes of
anti-anti-MHC antibodies
(internal image of anti-allo MHC; may
interact with self MHC)

GENERATION OF ANTI-IDIOTYPIC RECEPTORS TO ANTI-MHC RECEPTORS

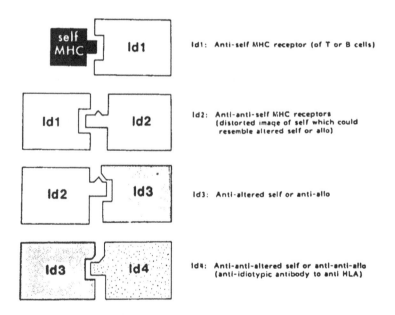

Id1: Anti-self MHC receptor (of T or B cells)

Id2: Anti-anti-self MHC receptors
(distorted image of self which could
resemble altered self or allo)

Id3: Anti-altered self or anti-allo

Id4: Anti-anti-altered self or anti-anti-allo
(anti-idiotypic antibody to anti HLA)

FIGURE 1. Generation of anti-MHC autoantibodies.

A possible explanation for the development of natural auto- and/or alloantibodies to HLA antigens is that allosensitization of immunosuppressed individuals through transplantation leads to the aberrant proliferation of B-cell clones producing antibodies against self-MHC antigens (Ab_1). Such Ab_1 anti-HLA could then trigger the development of Ab_2 representing an altered image of self. This distorted internal image may resemble allogeneic MHC antigens and stimulate the production of Ab_3, i.e., of anti-allo-MHC antibodies (Figure 1).

The reverse cycle of events is also possible. The patient may first develop anti-allo-MHC antibodies (Ab_1) to the graft. Anti-Id antibodies (Ab_2) triggered by antigraft-MHC may represent the distorted image of allo-MHC and stimulate the development of Ab_3 reacting with self or with irrelevant MHC antigens. Whatever the explanation, the fact remains that the organism maintains the capacity of producing anti-HLA autoantibodies and that antiself-reactive cells are not deleted from the immune repertoire.

It is possible that anti-HLA autoantibodies play a role in the pathogenesis of certain autoimmune diseases, in which alterations of self-MHC antigens have been implicated. It is also conceivable that such autoantibodies are involved in acquired immune deficiency syndrome (AIDS), a disease predominantly encountered in homosexuals and drug addicts.[18]

The presence of antilymphocytic autoantibodies has long been recognized as a characteristic feature in AIDS. Male homosexuals are highly sensitized through exposure to allogeneic sperm and develop anti-HLA and anti-anti-HLA antibodies. The latter may include imperfect images of self-MHC. The complementary anti-Id, consisting of anti-HLA autoantibodies, could prevent T-cell recognition of self-HLA plus X complexes, thus, accounting for impaired T-cell function. Furthermore, in the presence of complement, such autoantibodies might produce lymphocytolysis. Although this could explain the significant depletion of at least one subset of T lymphocytes (the T_4-positive cells), it does not explain why other cells are not also affected. It seems more likely that the depletion of T_4-positive cells is the result of a complex chain of events involving autoantibodies directed against T_4, the antigen which marks the cells that harbor the virus.

VI. ANTI-IDIOTYPIC T-CELL RESPONSES AND THEIR IMPLICATION IN AUTOIMMUNE DISEASES AND IN AIDS

It has long been suspected that, similar to B lymphocytes, T cells bear idiotypic receptors for antigens, and that T-B interaction is mediated through the complementarity of Id. A large body of literature supports this view.[9-16]

Our own studies have shown that T lymphocytes are activated in the autologous mixed lymphocyte reaction (AMLC) by Id expressed on autologous antigen-specific T-cell clones. We demonstrated the anti-idiotypic nature of the AMLC response by showing that the memory of this immune response is directed against the receptors expressed by the antigen-primed T cells. Such anti-idiotypic T-cell clones generally act as suppressors of "idiotypic" antigen-specific clones.[48-53]

If anti-idiotypic T cells act as suppressors of idiotypic alloreactive lymphocytes, there is no need to postulate the existence of "suppressor" cells as a distinct functional category, but rather a generation of cells with receptors that are complementary to the receptors of antigen-responsive cells.[42] This model may explain recent data on the molecular structure of the T-cell antigen receptor. Indeed, to our knowledge no convincing evidence has yet been found that the receptor of suppressor cells, whose function is to react against and to inhibit other "self" components of the immune response, differs from the receptor of cells reacting with the nominal antigen.[39,40]

The mechanism used by anti-idiotypic T cells to medicate suppression of helper, amplifier, or cytotoxic T cells is unknown. It is possible that such cells produce "suppressor" factors, or that they directly kill their target-idiotypic T-cell donor. It is also conceivable that such anti-idiotypic suppressor cells operate through a mechanism of help, or by activating B cells which produce anti-T-cell-receptor autoantibodies.[42] Support for this hypothesis comes from recent data in our laboratory. We have generated AMLC-reactive T-cell clones by activating T cells from individual A in MLC with autologous blasts primed to B. When added to MLC in which fresh cells from A were stimulated with irradiated lymphocytes from B, these clones acted as suppressors, i.e., they inhibited the blastogenic response by more than 50%, even when used at very low concentrations (500 to 1000 clones cells per 5×10^4 responding and 5×10^4 stimulating cells).[42] Supernatant from such cultures, but not from control cultures, contained IgG antibodies which reacted with 5 to 8% of resting T lymphocytes from A, but not with lymphocytes from other individuals. These antibodies may represent anti-T-cell-receptor autoantibodies which account for the suppression of MLC.

Our data, therefore, suggest that suppression of T-cell-immune responses to alloantigens is mediated by T cells either via the interaction of idiotypic and anti-idiotypic T-cell clones or through the mediation of B cells which form anti-T-cell-receptor autoantibodies.[42] A breakdown of these mechanisms of suppression may be involved in certain autoimmune diseases.

A variety of autoimmune conditions have been shown to display linkage disequilibrium to HLA, and the hypothesis that this might be due to the association of certain HLA genes with particular immune disease genes has been proposed.[1,2]

HLA-linked immune response genes do not explain, however, why the disease develops in some members of a family and not in others, particularly in the case of identical pairs of twins who are discordant for the disease. Environmental factors such as viruses may be involved and the disease may depend upon exposure to such factors.

Since monozygotic twins are identical for all genes except for those which undergo somatic mutations or rearrangements, as is the case for immunoglobulin and T-cell-receptor genes, we postulate that the difference between affected and nonaffected individuals depends on the set of V-region genes which are expressed.

It is conceivable that anti-idiotypic and idiotypic receptors use different families of V genes. Such a situation has been recently described in the case of antirheumatoid factor antibodies which use, almost invariably, the same family of V genes.[54] By analogy, in HLA-associated autoimmune diseases, the HLA-linked susceptibility gene may be the same in all individuals who have inherited the "disease" haplotype, yet what differs could be their capacity of autoinhibiting the immune response against the target organ. The nonaffected individuals may develop anti-T-cell-receptor antibodies or anti-idiotypic T suppressor cells, by utilizing a certain family of V genes required for the generation of the anti-idiotypic component. The affected individuals may not express or utilize these genes, so that the process of target organ destruction proceeds uninhibited. Such individuals are then more prone to develop antiself-MHC reactions, by lack of anti-antiself-MHC responses. Support for this hypothesis comes from our recent finding that patients with JDM or SLE have an increased frequency of anti-HLA autoantibodies. It, therefore, appears that the natural mechanism of down regulation of antiself-responses is impaired in these populations.

Anti-idiotypic T-cell responses may also be implicated in the horizontal transmission of HTLV III within the population of T4-positive lymphocytes. Although viral infection severely impairs T-cell-immune responses, the frequency of infected cells in AIDS patients is less than 1 in 10,000.

Viral cytotoxicity could be explained by the following mechanism. Monocytes may represent the first cells infected by the virus. Cells harboring the virus could express self-MHC plus viral complexes. Lymphocytes of the T4 subset, bearing "idiotypic receptors" for such self- plus X (HTLV III-product) antigens, subsequently recognize and proliferate clonally in response to this neoantigen. Proliferating cells, trying to attack and kill the target lymphocytes, would express an activation antigen which represents the receptor for the virus. This would allow the contamination of the effector cell by its virus-infected target.

The virus-infected T-cell clone might, in turn, become susceptible to the cytopathic effect of the virus or it might become the target of other cells which recognize self-MHC-plus-virus complexes or which are anti-idiotypic to the former, i.e., which recognize the T-cell antigen receptor on the surface of the first generation of infected cells. This chain of events, consisting of unsuccessful attempts of killer cells to destroy their target or in a cascade of idiotypic-anti-idiotypic T-cell responses, might perpetuate the infection. The clonality of such reactions may explain the low frequency of HTLV III-infected cells in the circulation. If cells participating in these events are "autoregulatory" T-cell clones, they may also induce autoantibodies aimed to react with virus-infected T cells. Cross reactions with normal lymphocytes could then result in the severe lymphocytopenia and immune deficiency which characterize the disease.

VII. CONCLUSIONS

Several years ago we proposed the concept that autoimmune anti-idiotypic responses to HLA are a necessity for the survival of the species, since they secure maternal tolerance to the fetus and, by implication, the preservation of MHC polymorphism.

Our argument was based on paradoxical immune phenomena occurring spontaneously in nature.

We emphasized the conceptual conflict between admitting that selective pressure has operated in the generation of the extreme polymorphism of MHC and that, at the same time, MHC recognition has evolved as a mechanism of self-nonself-discrimination which drives the somatic diversification of the T-cell receptor repertoire. Indeed, these two mechanisms cannot be easily reconciled, since recognition of nonself should lead to rejection of the semiallogenic concept and, thus, to selection against rather than for polymorphism of MHC genes.[18,19]

Since MHC heterozygocity and not homozygocity is at advantage in random populations, it appears that mechanisms suppressing maternal alloimmunity against the fetus must be at play.

Down regulation of alloimmune responses may occur if Id expressed by anti-HLA antibodies and/or by alloreactive T-cell clones trigger a second wave of anti-Id antibodies or anti-idiotypic T cells which block the first idiotypic generation.

The potentials for both positive and negative controls mediated through network interactions are enormous, not only for responses to foreign antigens, but also in the generation and diversification of the immune repertoire due to the internal selective pressure provided by Id.

In this perspective one can conceive a scheme in which the enormous complexity of MHC has been preserved, because MHC-products represent the starting point of the self-recognition process. Rather than being deleted during intrathymic differentiation, as proposed by Jerne,[7] cells with receptors for self may represent the starting point in the generation of sequential waves of immunocytes with anti-antiself-receptors (Figure 1).

The antigen combining sites expressed by the first wave of cells bearing antiself-MHC receptors (G1 generation) are complementary to the antigen (self-MHC) representing the negative or antiimage of the (self-MHC) antigen. They can induce a second wave of cells with receptors (G2) which are complementary to the binding site of the antiimage and, thus, represent the internal image of (self) MHC antigens. Imperfect internal images may "look like" nonself MHC.

Anti-idiotypic T and B lymphocytes (G2) may act as suppressors of antiself-reactive clones. They might also diversify their receptors by rapid mutational events (as proposed by Jerne) for antiself rather than anti-antiself-receptors. G2 (anti-idiotypic) receptors, which represent "distorted" internal images of self, resembling allo-MHC, may trigger a third wave (G3) with high affinity for allelic variants of self-MHC antigens. These receptor models may account for the high frequency of alloreactive T cells which are present in man and animals prior to any known sensitization to non-self MHC. If G3 receptors are built into the repertoire, G4 receptors which should be anti-idiotypic to such "allo-like" receptors could provide the pool from which suppressors of alloimmune reactions will emerge.

We have deliberately refrained from specifying whether we refer to the repertoire of T-cell receptors or of antibody molecules, since we believe that the driving force that sets the necessary size of the immune vocabulary may reside in idiotypic determinants of B- as well as T-cell receptors. Recent advances in the understanding of the structure of the T-cell receptor substantiate, indeed, the idea that the generation of diversity resides, as in the case of antibodies, in the choice of germ-like genes, mutations, somatic V-D-J rearrangements, and recombinations.

Our model integrates, however, the response to self- and allo-MHC in the mechanism of normal autorecognition and assumes that normalization of the immune repertoire occurs through cellular selection.

ACKNOWLEDGMENT

This work is supported by N.I.H. grants A125210-01 and HL36581. The authors greatfully acknowledge the contribution of Dr. Jay Lefkowitz, Associate Professor of Pathology, College of Physicians and Surgeons of Columbia University, who provided the illustrations.

REFERENCES

1. **Suciu-Foca, N.,** The HLA system in human pathology, *Pathobiol. Ann.,* 9, 81, 1979.
2. **Suciu-Foca, N. and King, D. W.,** The HLA system and the immune response in man, *Immunological Diseases,* 4th ed., in press.
3. **Hood, L. E., Weissman, I. L., Wood, W. B., and Wilson, J. H.,** *Immunology,* 2nd ed., Benjamin/Cummings, Menlo Park, Calif., 1984.
4. **Finberg, R., Burakoff, S. T., Cantor, H., and Benacerrat, B.,** Biological significance of alloreactivity: T cells stimulated by Sendai virum-coated syngeneic cells specifically lyse allogeneic target cells, *Proc. Natl. Acad. Sci. U.S.A.,* 75, 5145, 1976.
5. **Srendni, B. and Schwartz, R. M.,** Alloreactivity of an antigen specific T cell clone, *Nature,* 287, 855, 1980.
6. **Braciale, T. J., Andrew, M. E., and Braciale, N. L.,** Simultaneous expression of H-2-restricted and alloreactive recognition by a cloned line of influenza virus-specific cytotoxic T lymphocytes, *J. Exp. Med.,* 153, 1271, 1981.
7. **Jerne, H. D.,** Towards a network theory of the immune system, *Ann. Immunol. (Paris),* 125C, 373, 1974.
8. **Jerne, H. D.,** Idiotypic networks and other preconceived ideas, *Immunol. Rev.,* 79, 5, 1984.
9. **Bona, C. A., Victor-Kohvin, C., and Manheimer, A. J.,** Regulatory arms of the immune network, *Immunol. Rev.,* 79, 25, 1984.
10. **Cooper, J., Eichmann, K., Fey, K., Melchers, I., Simon, M. M., and Weelzein, H. A.,** Network regulation among T cells. Qualitative and quantitative studies on suppression in the non-immune state, *Immunol. Rev.,* 79, 63, 1984.
11. **McNamara, M., Gleason, K., and Kohler, H.,** T cell helper circuits, *Immunol. Rev.,* 79, 87, 1984.
12. **Takemori, T. and Rajewsky, J.,** Mechanism of neonatally induced idiotype suppression and its relevance for the acquisition of self tolerance, *Immunol. Rev.,* 79, 103, 1984.
13. **Bona, C. A. and Kohler, H.,** Immune networks, *Ann. N.Y. Acad. Sci.,* 418, 1983.
14. **Bona, C. A.,** Idiotypes and lymphocytes, in *Immunology,* Dixon, F. J. and Kunkel, A., Eds., Academic Press, New York, 1981.
15. **Koprowski, H. and Melcher, F.,** *Current Topics in Microbiology 119 and Immunology,* Springer-Verlag, Berlin, 1985.
16. **Bluestone, J. A., Leo, O., Epstein, S. L., and Sachs, D. H.,** Anti-idiotypic antibodies as immunogens, *Immunol. Rev.,* 90, 5, 1986.
17. **Gill, T. J.,** Immunogenetics of spontaneous abortions in humans, *Transplantation,* 35, 1, 1983.
18. **Suciu-Foca, N., Kohler, H., and King, D. W.,** Anti-idiotypic autoimmunity — A necessity for species survival, *Surv. Immunol. Res.,* 3, 311, 1984.
19. **Suciu-Foca, N., King, D. W., Reemstma, K., and Kohler, H.,** Autoimmunity and self antigens, *Concepts Immunopathol.,* 1, 173, 1984.
20. **Suciu-Foca, N., Rohowsky-Kochan, C., Reed, E., Haars, R., Bonagura, V., King, D. W., and Reemstma, K.,** Idiotypic network regulations of immune responses to HLA, *Fed. Proc. Fed. Am. Soc. Exp. Biol.,* 44, 2483, 1985.
21. **Suciu-Foca, N., Reed, E., Rohowsky, C., Kung, P., and King, D. W.,** Anti-idiotypic antibodies to anti-HLA receptors induced by pregnancy, *Proc. Natl. Acad. Sci. U.S.A.,* 80, 830, 1983.

22. **Reed, E., Bonagura, V., Kung, P., King, D. W., and Suciu-Foca, N.,** Anti-idiotypic antibodies to HLA-DR4 and DR2, *J. Immunol.,* 131(6), 2890, 1983.

23. **Bonagura, V., Rohowsky-Kochan, C., Reed, E., Ma, A., and Suciu-Foca, N.,** Perturbation of immune network in herpes gestations, *Hum. Immunol.,* 14, 211, 1986.

24. **Suciu-Foca, N., Reed, E., Rohowsky-Kochan, C., Popovic, M., Bonagura, V., King, D. W., and Reemstma, K.,** Idiotypic network regulations of the immune response to HLA, *Transplant. Proc.,* 17, 716, 1985.

25. **Reed, E., Rohowsky-Kochan, C., and Suciu-Foca, N.,** Analysis of 9W antisera detecting DR4 and DR2 associated epitopes by use of anti-idiotypic antibodies, in *Histocompatibility Testing,* Albert, E. D., Baur, M. P., and Mayr, W. R., Eds., Springer-Verlag, New York, 1984, 422.

26. **Binz, H. and Wigzel, H.,** Shared idiotypic determinants on B and T lymphocytes reactive against the same antigenic determinants, *J. Exp. Med.,* 142, 197, 1975.

27. **Anderson, L. C., Binz, H., and Wigzel, H.,** Specific unresponsiveness to transplantation antigens induced by auto-immunization with syngeneic, antigen-specific T lymphoblasts, *Nature,* 264, 778, 1976.

28. **Acuto, O., Hussey, R., Fitzgerald, K., Protentis, J., Meur, S., Schlossman, S., and Reinherz, E.,** The human T cell receptor appearance in ontogeny of biochemical relationship of and subunits IL_2 dependent clones and T cell tumors, *Cell,* 34, 717, 1983.

29. **McIntyre, B. and Allison, J.,** The mouse T cell receptor: structural heterogeneity of molecules of normal T cells defined by xenoantiserum, *Cell,* 34, 739, 1983.

30. **Haars, R., Rohowsky-Kochan, C., Reed, E., King, D. W., and Suciu-Foca, N.,** Modulation of T cell-antigen receptor on lymphocyte membrane, *Immunogenetics,* 20, 397, 1984.

31. **Aleksander, I. and Mak, T. W.,** A human T cell-specific cDNA clone encodes a protein having extensive homology to immunoglobulin chains, *Nature,* 308, 145, 1984.

32. **Hedrick, S., Cohen, D., Nielson, E., and Davis, M.,** Isolation of cDNA clones encoding T cell specific membrane associated proteins, *Nature,* 308, 149, 1984a.

33. **Hedrick, S., Nielsen, E., Kabaler, J., Cohen, D., and Davis, M.,** Sequence relationships between putative T cell receptor polypeptides and immunoglobulins, *Nature,* 308, 153, 1984b.

34. **Saito, H., Kranz, D. M., Takagaki, Yl., Hayday, A. C., Eisen, J. N., and Tonegawa, S.,** Complete primary structure of a heterodimeric T cell receptor derived from cDNA sequences, *Nature,* 309, 757, 1984.

35. **Saitor, H., Kranz, D. M., Takagaki, V., Hayday, A. C., Eisen, H., and Tonegawa, S.,** A third rearranged and expressed gene in a clone of cytotoxic T lymphocytes, *Nature,* 312, 36, 1984.

36. **Mac, T. and Yanagi, Y.,** Genes encoding the human T cell antigen receptor, *Immunol. Rev.,* 81, 221, 1984.

37. **Chien, Y., Becker, D. M., Lindsten, T., Ocamura, M., Cohn, D., and Davis, M.,** A third type of murine T cell receptor gene, *Nature,* 312, 31, 1984.

38. **Hayday, A. C., Saito, H., Gillies, S. D., Kranz, D. M., Tanigawa, G., Eisen, H. N., and Tonegawa, S.,** Structure, organization and somatic rearrangement of T cell gamma genes, *Cell,* 40, 259, 1985.

39. **Toyonaga, B., Yanagi, Y., Suciu-Fona, N., Minden, M. D., and Mak, T. W.,** Rearrangement of the T cell receptor gene YT3J in human DNA from thymic leukemic T cell lines and functional helper, killer, and suppressor T cell clones, *Nature,* 311, 385, 1984.

40. **Yoshikai, Y., Yanagi, Y., Suciu-Foca, N., and Mak, T. W.,** Presence of T cell receptor mRNA in functionally distinct T cell and elevations during intrathymic differentiation, *Nature,* 310, 506, 1984.

41. **Kronenberg, M., Governman, J., Haars, R., Mallisen, M., Kraig, E., Phillips, L., Delovitch, T., Suciu-Foca, N., and Hood, L.,** Rearrangement and transcription of the B chain genes of the T cell antigen receptor in different types of murine lymphocytes, *Nature,* 313, 647, 1985.

42. **Suciu-Foca, N., Reemstma, K., and King, D. W.,** The significance of the idiotypic anti-idiotypic network in man, *Transplant. Proc.,* 18, 230, 1986.

43. **Mowbray, J. F., Liddell, H., Underwood, J. L., Gibbings, C., Reginald, P. W., and Berard, P. W.,** Controlled trial of treatment of recurrent spontaneous abortion by immunization with paternal cells, *Lancet,* p. 941, 1985.

44. **Reed, E., Hardy, M., Lattes, C., Brensilver, J., McCabe, R., Reemstma, K., Suciu-Foca, N.,** Anti-idiotypic antibodies and their relevance to transplantation, *Transplant. Proc.,* 17, 7350, 1985.

45. **Gocken, N. E. and Clinical Affairs Committee,** Outcome of renal transplantation following a positive cross-match with historical sera: the ASHI survey, *Hum. Immunol.,* 14, 77, 1985.

46. **Ivanyi, P., Van Mourik, P., Brenning, M., Pruisbeek, A. M., and Krose, C. J. M.,** Natural H2-specific antibodies in sera of aged mice, *Immunogenetics,* 15, 95, 1982.

47. **Tongio, M. M., Falkenrodt, A., Mitsuichi, Y., Urlacher, A., Bergerat, J. P., North, M. L., and Mayer, S.,** Natural HLA antibodies, *Tissue Antigens,* 26, 271, 1985.

48. **Suciu-Foca, N., Rohowsky, C., Kung, P., and King, D. W.,** Idiotypic-like determinants on human T lymphocytes alloactivated in mixed lymphocyte culture (MLC), *J. Exp. Med.,* 156, 283, 1982.

49. **Suciu-Foca, N., Rohowsky, C., Kung, A., Lewison, A., Nicholson, J., Reemstma, K., and King, D. W.,** MHC specific idiotypes on alloactivated human T cells: in vivo and in vitro studies, *Transplant. Proc.,* 15(1), 784, 1983.

50. **Suciu-Foca, N., Rohowsky, C., Coburn, C., Reed, E., Khan, R., and Lewison, A.,** Alloreactive T cells: expression of HLA-D antigens, stimulation of autologous MLR and possible immunoregulatory function, *Hum. Immunol.,* 3, 301, 1981.

51. **Suciu-Foca, N., Rohowsky, C., Kung, P., and King, D. W.,** Idiotypic receptors for soluble antigens on human T lymphocyte clones, *Hum. Immunol.,* 9, 34, 1984.

52. **Rohowsky-Kochan, C., Kung, P., King, D. W., and Suciu-Foca, N.,** Enhancement of idiotype-anti-idiotype T cell interactions by monoclonal OKT11A antibody, *Hum. Immunol.,* 9, 103, 1984.

53. **Rohowsky, C., Suciu-Foca, N., Kung, P., Tang, T. F., Reed, E., and King, D. W.,** Suppressor T cells generated in autologous MLR with MLC activated T lymphoblasts, *Transplant. Proc.,* 15(1), 765, 1983.

54. **Rohowsky, C., Reed, E., Suciu-Foca, N., Kung, P., Reemstma, K., and King, D. W.,** Inhibition of MLC reactivity to autologous allo-activated T lymphoblasts by sera from renal allograft recipients, *Transplant. Proc.,* 15(3), 1761, 1983.

55. **Victor-Kobrin, C., Manser, T., Moran, T. M., Imanishi-Kari, T., Gefter, M., and Bona, C. A.,** Shared idiotopes among antibodies encoded by heavy chain variable region (VH) gene members of the J558 VH family as basis for cross-reactive regulation of clones with different antigen specificity, *Proc. Natl. Acad. Sci. U.S.A.,* 82, 7696, 1985.

56. **Bonaugura, V., Averil, Ma., McDowell, J., Lewison, A., King, D. W., and Suciu-Foca, N.,** Anti-clonotypic autoantibodies in pregnancy, *Cellular Immunol.,* 108, 356, 1987.

Chapter 11

THE REGULATION OF THE AGING IMMUNE RESPONSE BY AUTO-ANTI-IDIOTYPIC ANTIBODY*

E. A. Goidl

TABLE OF CONTENTS

* Supported in part by USPHS grant AG 04042.

I. INTRODUCTION

In contrast to the decline of most physiological functions which occur in aging, there is an increase in the production of auto-anti-idiotypic antibody (auto-anti-Id) in the senescent immune response. As proposed by Jerne,[1] the immune response may be represented as an interacting network of antibody molecules. Simply, this network comprises antiantigen antibodies (Id antibodies or Ab_1) and anti-anti-antibodies (anti-Id or Ab_2). This network of interacting elements is disturbed from its steady state by the introduction of antigen, and an immune response is generated. Following the production of Id and anti-Id antibodies, all elements then achieve a new steady state.

This steady state has been looked at as "positive auto-immunity" by Wigzell.[2] This is to be contrasted by "negative autoimmunity" best exemplified by autoimmune diseases in which a number of autoantibodies (Ab_1) are produced against autologous determinants. In this chapter, I shall try to show how this balance between positive and negative autoimmunity changes during the life of the animal and is regulated by the production of auto-anti-Id. I shall also produce evidence that changes in positive autoimmunity may be required to control the alterations seen in negative autoimmunity with respect to increase of age.

We have shown extensively the changes in the production of auto-anti-Id antibody with aging of the animal. Using a murine experimental system, we have provided evidence that there occurs a progressive relative increase of auto-anti-Id production during the normal response of aged mice.

I shall review the experiments which have led us to dissect the cellular control of the increased production of auto-anti-Id antibody seen in the normal immune responses of aging mice.

II. MAGNITUDE AND CHARACTERISTICS OF THE AGED IMMUNE RESPONSE

In aging there is a decrease in the magnitude of the immune response of aged mice which is accompanied by a preferential loss of high affinity antibody and by a decrease in the heterogeneity of antibody affinity.[3] When aged BALB/c and C57L/J mice of different ages were immunized with 500 μg of dinitrophenylated bovine gamma globulin (DNP-BGG), the magnitude of the immune response of aged mice was considerably lower than that of young mice (see Table 1). The immune response of young mice is characterized by a maturation process which leads to an increase in average affinity and an increase in heterogeneity of antibody affinity with respect to the time after antigen administration. In contrast, the immune response of aged mice is characterized by a decreased average affinity of the antibody produced and also by a dramatic decrease in the heterogeneity of antibody affinity produced. This decreased heterogeneity of antibody affinity is preferentially seen in the loss of high affinity antibody production. The aged immune response by its decreased magnitude and by its relative loss of high affinity antibody may contribute to the increased susceptibility to infectious agents seen in aging.

III. PRODUCTION OF AUTO-ANTI-ID ANTIBODY IN THE IMMUNE RESPONSE OF AGED MICE

The decrease in the immune response which occurs in aging has been attributed to a decline in both T- and B-cell functions. We shall look at the evidence that increased production of auto-anti-Id in aging immunity may effectively lead to a decrease in the magnitude of the response seen in aging.

We have documented the production of auto-anti-Id during the course of the normal immune response by the use of two different methods.

Table 1
COMPARISON OF THE IMMUNE RESPONSES OF YOUNG AND AGED MICE

Strain	Age (months)	Direct PFC	Indirect PFC	Heterogeneity index
C57L/J	2	2,500	8,000	2.8
	12	3,200	6,400	2.3
	25	900	1,200	1.6
BALB/c	2	9,600	11,000	2.8
	24	1,900	2,000	1.8

Note: Mice were immunized with DNP-BGG in complete Freund's adjuvant and sacrificed 2 weeks later. Heterogeneity index with respect to avidity is based on the Shannon function.

Adapted from Goidl, E. A., Innes, J. B., and Weksler, M. E., *J. Exp. Med.,* 144, 1037, 1976.

1. We have shown that in the presence of low concentrations of free hapten there frequently occurs an increase in the number of plaque-forming cells (PFC) enumerated in a Jerne localized-in-gel hemolytic PFC assay.[4] These potential PFC which are hapten augmentable represent cells whose secretion of antibody has been blocked by the binding of auto-anti-Id to cell-surface idiotypic determinants. In this system, we view that the hapten may essentially compete with anti-Id for binding with cell surface antibody molecules. In a very narrow window of affinity distributions involving both the affinities of Id antibody for the hapten and the respective affinities of Id and anti-Id antibodies, hapten is able to displace bound anti-Id antibody, and subsequent to this event, inhibition of secretion is reversed. We have also shown that the presence of auto-anti-Id in the serum can be assayed by its capacity to inhibit PFC formation in a hapten-reversible fashion;[5,6]

2. We have also developed an enzyme-linked immunosorbent-assay (ELISA) for auto-anti-Id antibody.[7] This assay utilizes an enzyme-labeled, affinity-purified IgM from immune animals as the Id probe for anti-Id. This assay can detect both heterologous and autologous anti-Id. Results obtained by this assay at the serological level have confirmed those seen at the cellular level using hapten-augmentable PFC as a measure of the production of auto-anti-Id.

These methods were used to determine the extent of the contributions of the immune network to the regulation of the immune response of aged mice. Several inbred strains of mice (C57BL/6, BALB/c, CBF₁, [C57BL/6 × BALB/c]F₁, and AKR/J) of ages ranging from 6 to 100 weeks of age were immunized with 10 μg of trinitrophenylated-lysyl-ficoll (TNP-F). The magnitude of the anti-TNP PFC response as well as the percentage of hapten-augmentable PFC was determined at several times following antigen administration.[8] These studies (Table 2) showed that there is a relative increase in the auto-anti-Id antibody response with age. This production contributes to a greater down regulation of the immune response of the aged animals. The increase in the production of auto-anti-Id in the aging immune response could be due to the recognition of a larger proportion of anti-TNP idiotopes or, alternatively, a reduced heterogeneity of clonotypes responding to the antigenic determinant assay could contribute to this phenomenon. It is not possible at this time to determine which of these two mechanisms is responsible for the increased production of auto-anti-Id in aging. Szewczuk and Campbell[9] also reported the increase of auto-anti-Id production in the aged immune response to thymic-dependent antigens.

Table 2
EFFECT OF AGE ON PRODUCTION OF AUTO-ANTI-ID ANTIBODY

Strain	Age (months)	Overall average of hapten-augmentable PFC (%) (x = SEM)
C57BL/6	2	39 ± 16
	3	12 ± 9
	20	532 ± 256
BALB/c	2	15 ± 6
	7	37 ± 20
	25	49 ± 25
AKR/J	2	17 ± 6
	6—7	110 ± 21
CBF₁	2—3	20 ± 6
	20	190 ± 94

Note: Mice were immunized with 10 μg TNP-F intravenously and sacrificed 7 days later.

Adapted from Goidl, E. A., Thorbecke, G. J., Weksler, M. E., and Siskind, G. W., *Proc. Natl. Acad. Sci. U.S.A.*, 77, 6788, 1980.

Table 3
CHANGES IN ID EXPRESSION WITH AGE

| Source of anti-TNP PFC[a] | Incidence of suppression (%); age of source of anti-Id | | |
	3—4 weeks old	6—8 weeks old	21—22 months old
3—4 weeks	58	15	20
6—8 weeks	0	67	50
21—22 months	10	7	81

Note: Italicized values reflect the highest correspondence between Id and anti-Id.

[a]   C57BL/6 mice were immunized by the i.v. injection of 10 μg TNP-F.

Adapted from Goidl, E. A., Thorbecke, G. J., Weksler, M. E., and Siskind, G. W., *Proc. Natl. Acad. Sci. U.S.A.*, 77, 6788, 1980.

The effect of age on Id expression was also determined.[8] We used hapten-reversible inhibition of PFC formation as an assay for anti-Id antibody and Id antibody-secreting cells. Sera from aged (21- to 22-month-old) C57BL/6 mice immunized with TNP-F significantly inhibited plaque formation in a hapten-reversible manner by spleen cells obtained from 81% of TNP-F immunized age-matched mice (Table 3). When the same sera were tested on Id-bearing spleen cells obtained from TNP-immune 6- to 8- or 3- to 4-week-old mice, a different pattern emerged. Sera from aged animals inhibited plaque formation from only 50% of 6- to 8-week-old mice and 20% of immature (3- to 4-week-old) mice. Remarkably, sera from TNP-F immunized mice inhibited preferentially idiotypic-bearing lymphocytes of the same age as the mice from whom the sera were obtained. This was also reinforced by the fact that sera from a given age group rarely inhibited plaque formation by cells from mice of other age groups. These data strongly support the conclusion that with increasing age there

Table 4
THE INCREASED AUTO-ANTI-ID ANTIBODY
RESPONSE IS A TRAIT STABLE UPON
SPLEEN CELL TRANSFER

Age of donors (months)	Direct anti-TNP PFC (PFC/spleen ± SE)	Hapten-augmentable PFC (%) (x ± SE)
4	17,000 ± 3,000	6 ± 5
18	10,000 ± 1,200	109 ± 25

Note: Lethally irradiated mice were reconstituted with spleen cells from syngeneic donors of the indicated age. Recipients were immunized with 10 μg TNP-F administered intravenously and their response assayed 7 days later.

Adapted from Goidl, E. A., Choy, J. W., Gibbons, J. J., Weksler, M. E., Thorbecke, G. J., and Siskind, G. W., *J. Exp. Med.*, 157, 1635, 1983.

is a gradual change in the catalog of Id expressed after stimulation by antigen. In this case the TNP determinant was presented in a thymic-independent carrier. These data also seem to make it unlikely that Id expression is purely a random event. This is shown by the preferential auto-anti-Id and Id recognition within the same groups. Also noteworthy is the finding that sera from one age group do not recognize the full repertoire of anti-TNP Id produced by age-matched immune spleen cells, indicating that syngeneic mice of a given age differ to a certain extent in Id expression.

The shift in clonotypic antibody production may reflect changes in the distribution and/or expression of clonotypes as a consequence of repeated exposure to common antigens and infectious agents in the environment. This shift in Id antibody repertoire with age was only analyzed at the level of expression; these results do not imply that such changes stem from differences in the catalog of Idiotypes present at the genomic level.

The most dramatically affected subpopulations of PFC down regulated by auto-anti-Id are high affinity antibody-secreting cells. This is seen more readily, perhaps, because of the limitations of the assay system, but, nevertheless, this down regulation in high affinity antibody may contribute to the increased susceptibility to infectious agents seen in aging.

In conclusion, the relative increase of the production of auto-anti-Id with increasing age is a general phenomenon as shown by elevated responses in aged mice of different inbred strains. Also, with respect to age, there occurs a gradual shift in the expression of the repertoire of Id antibody. The precise nature of this shift has not been determined.

IV. CELLULAR REGULATION OF AUTO-ANTI-ID ANTIBODY PRODUCTION IN THE IMMUNE RESPONSE OF AGED MICE

The production of auto-anti-Id has been shown to be highly thymic dependent. In this section we shall describe experiments leading to the identification of the cell regulating the production of auto-anti-Id in the aging immune response. To determine the cellular regulation of the production of auto-anti-Id in the aged immune response, we first established whether this increase was a trait which was stable upon cell transfer.[10] Spleen cells were obtained from young or aged C57BL/6 mice. Cells were transferred along with 10 μg TNP-F into lethally irradiated syngeneic recipients. Results (Table 4) showed that mice reconstituted with spleen cells from aged donors produced high levels of hapten-augmentable PFC in

Table 5
BONE MARROW CELLS FROM AGED MICE ARE
"YOUNG-LIKE" WITH RESPECT TO LEVELS OF AUTO-
ANTI-ID PRODUCTION

Age of bone marrow[a] donors (months)	Direct anti-TNP PFC/ spleen (x̄ ± SE)	Hapten-augmentable PFC (%) (x̄ ± SE)
2—3	4000 ± 600	13 ± 3
18	1400 ± 200	7 ± 5

[a] 5×10^7 Bone marrow cells from donors of indicated ages were transferred into lethally irradiated syngeneic 2- to 3-month-old mice, along with 10 μg TNP-F. Assays were performed 7 days later.

Adapted from Goidl, E. A., Choy, J. W., Gibbons, J. J., Weksler, M. E., Thorbecke, G. J., and Siskind, G. W., *J. Exp. Med.*, 157, 1635, 1983.

contrast to those mice which had received cells from young donors, which had low levels of hapten-augmentable PFC. Thus, the high percentage of hapten-augmentable PFC which had previously been reported to be a characteristic of the immune response of aged mice[8] is also seen in those animals which had been reconstituted with spleen cells from aged donors. The relative increase of auto-anti-Id production seen in the immune response of aged mice is, therefore, a trait which is stable upon cell transfer to syngeneic lethally irradiated recipients and is not a phenomenon dependent upon the internal physiological milieu of the aged mouse.

The increased production of auto-anti-Id in the aged immune response[8,9] had so far been a characteristic of the peripheral lymphoid system (i.e., spleen). We next decided to determine whether this characteristic extended to the lymphoid cells obtained from the central lymphoid organs (i.e., bone marrow) of young and aged mice. In a series of experiments (Table 5) 5 × 10^7 bone marrow cells obtained from young or aged mice were transferred into syngeneic lethally irradiated recipients. All recipients were immunized with 10 μg of TNP-F and 7 days later their responses were analyzed for the presence of antihapten PFC and for the production of hapten-augmentable PFC. Bone marrow cells were able to reconstitute recipients and mount a modest anti-TNP PFC response. In no case did we find the recipients producing any significant numbers of hapten-augmentable PFC. Lymphoid cells obtained from the bone marrow of either young or aged mice are, therefore, indistinguishable from each other, leading to the production of low to undetectable levels of auto-anti-Id antibody.

We had shown previously that T-cell-deficient animals when immunized with TNP-F fail to produce auto-anti-Id.[11] In these cases, data obtained from young adult mice showed that the production of auto-anti-Id is a thymic-dependent phenomenon. We next examined the thymic dependence of the production of auto-anti-Id in the aged immune response. Thymectomized, lethally irradiated syngeneic recipients were reconstituted with spleen cells from aged or young donor mice. Pools of cells from aged or young donors were treated with complement alone or with complement and monoclonal anti-Thy-1 antibody in order to eliminate T cells. Resultant cell populations were injected into recipients along with 10 μg TNP-F. The immune response of the recipients was analyzed for the presence of anti-TNP PFC and for hapten-augmentable PFC (Table 6). The results obtained showed that in contrast to the immune response of young mice (whose production of auto-anti-Id is a thymic-dependent event), with splenic non-T cells obtained from aged mice, the production of auto-anti-Id is a relatively thymic-independent phenomenon.

The experiments described so far point to the similarity of low levels of auto-anti-Id

Table 6
THE PRODUCTION OF AUTO-ANTI-ID IS
RELATIVELY THYMIC INDEPENDENT IN THE
SPLEEN OF AGED MICE

Treatment of cells	Direct anti-TNP PFC/spleen ($\bar{x} \pm$ SE)	Hapten-augmentable PFC (%) ($\bar{x} \pm$ SE)
Complement	1500 ± 250	42 ± 13
Anti-Thy 1.2 + complement	2000 ± 250	43 ± 13

Note: Lethally irradiated thymectomized C57BL/6 mice were reconstituted with spleen cells from 18-month-old mice. As indicated, spleen cells were treated with anti-Thy-1.2 and complement to remove T cells or with complement alone. Immunization has been described (see Table 5).

Adapted from Goidl, E. A., Choy, J. W., Gibbons, J. J., Weksler, M. E., Thorbecke, G. J., and Siskind, G. W., *J. Exp. Med.*, 157, 1635, 1983.

production between cells from bone marrow of young and aged mice. We also have documented a relatively low thymic dependency of auto-anti-Id production in the spleen of aged mice, while these cells give rise to high levels of auto-anti-Id antibody. Since lymphoid precursor cells arise in the bone marrow of adult animals and migrate to populate peripheral lymphoid organs, we next determined the identity of the cells responsible for the dramatic change in auto-anti-Id production seen between cells of the central and peripheral lymphoid organs of aged mice. Groups of lethally irradiated syngeneic young adult mice were reconstituted with cells from bone marrow of young or aged donors. It is important to note that, so far, all recipients were identical with respect to their capacity to produce auto-anti-Id. We have shown (Table 5) that bone marrow cells from aged or young donors are undistinguishable in this context. Next, some recipients received B-cell-depleted spleen cells from the spleen of either young or aged donors. The different mixture of cells was allowed to reside for a period of 5 days. At this time all recipients were immunized with 10 μg TNP-F and their immune response analyzed for anti-TNP PFC and for the presence of hapten-augmentable PFC, 7 days later.

Recipients of T cells from young donors had a low percentage of hapten-augmentable PFC, such as is typical of young mice, regardless of the source of the bone marrow cells (Table 7). In contrast, recipients of T cells from aged donors had a high incidence of hapten-augmentable PFC, as is typical of old mice, and this was the case whether these recipients had received bone marrow cells from either young or old donors. In both cases, the groups that received T cells from old donors had a significantly higher percentage of hapten-augmentable PFC than did the groups which had received T cells from the spleens of young donors. In conclusion, the magnitude of the production of auto-anti-Id antibody is regulated by the non-B splenic cell production. Since this population was shown to be >95% Thy-1⁺, it seems likely that the increased production of auto-anti-Id in the aged immune response is regulated by peripheral splenic T cells.

The data presented in this section so far attest to the complexity of events leading to the increased production of auto-anti-Id in the aged immune response. First we have shown that lymphocytes arising from central lymphoid organs of both young and aged mice are indistinguishable in their capacity to form auto-anti-Id. That these two populations lead to the same low auto-anti-Id productions is consistent with the findings of Ogden and Micklem[12] and Harrison et al.[13] that bone marrow cells from young and old are similar with respect to

Table 7
PERIPHERAL SPLENIC T CELLS CONTROL THE
PRODUCTION OF AUTO-ANTI-ID ANTIBODY

Age of cell donors (months)		Direct anti-TNP PFC/ spleen (x ± SE)	Hapten-augmentable PFC (%) (x ± SE)
Bone marrow cell donor	T-cell donor		
2—3	2—3	2300 ± 500	12 ± 4
2—3	18	1200 ± 200	55 ± 12
18	2—3	2300 ± 500	16 ± 5
18	18	1500 ± 600	38 ± 12

Note: Male C57BL/6 mice were lethally irradiated and reconstituted by the i.v. injection of 2.5 to 3.5 × 10⁷ bone marrow cells obtained from donors of indicated age and 2 to 4 × 10⁷ B cell-depleted spleen cells from donors of the indicated age. Recipients were immunized 5 days following cell transfer by the i.v. administration of 10 μg TNP-F. All assays performed 7 days following immunization.

Adapted from Goidl, E. A., Choy, J. W., Gibbons, J. J., Weksler, M. E., Thorbecke, G. J., and Siskind, G. W., *J. Exp. Med.*, 157, 1635, 1983.

stem cell activity. We have also shown that the increased production of auto-anti-Id in aging is a characteristic of the peripheral lymphoid populations and that the expression of this trait, which is transferrable, is under splenic T-cell control. Transferrability, of course, implies that this trait is not dependent on the internal milieu of the aged animal. Once this characteristic is acquired in the peripheral lymphoid system of the aged, it is, in contrast to young mice, relatively thymic independent. The dichotomy between the "young-like" bone marrow cells from old mice and the control exercised by the long-lived peripheral T cell allowed us to predict that if the peripheral lymphoid system of an old animal were acutely depleted of cells while the bone marrow was left intact (e.g., irradiation with bone marrow shielding) and the animal allowed to reconstitute its peripheral lymphoid system from its own marrow, it should behave like a young mouse with respect to the production of auto-anti-Id antibody. Furthermore, this type of manipulation should be alterable by the transfer into such irradiated animals of peripheral T cells from donors of different ages.

The experimental results obtained confirm these predictions. Mice which were exposed to a normally lethal dose of irradiation, while their bone marrow was partially shielded, survived and within 6 weeks regained normal immune function. At this time, animals were immunized with 10 μg TNP-F and their anti-TNP PFC and hapten-augmentable PFC and ELISA titers for auto-anti-Id determined 7 days following antigen administration. After recovery from irradiation, old mice produce low auto-anti-Id antibody (Table 8) comparable to the levels produced by young mice. This is true whether the production of auto-anti-Id is measured by the magnitude of hapten-augmentable PFC or by ELISA assay for serum auto-anti-Id from these animals. The magnitude of auto-anti-Id production is under the control of peripheral splenic T cells as demonstrated by T-cell transfer into such animals. When mice irradiated with their bone marrow shielded received splenic T cells from aged donors, they produced high levels of auto-anti-Id antibody; mice similarly irradiated which received splenic T cells from young donors produced low levels of auto-anti-Id antibody (Table 9).

These results are consistent with the hypothesis that the idiotypic catalog is similar in the bone marrow B cells of young and aged mice. These two populations are thought to be similar in their stem cell activity potential.[12,13] Furthermore, Zharhary and Klinman[14] demonstrated that the level of diversity of antibody repertoire for PR8 influenza virus is similar in young and old mice. The differences observed in the immune responses of young and

Table 8
CONCENTRATION OF AUTO-ANTI-ID IN THE SERUM OF MICE AFTER RECOVERY FROM IRRADIATION AND FOLLOWING IMMUNIZATION WITH TNP-F

Mice[a]		Binding of anti-TNP antibody
Age	Treatment	(A 405 nm)
18—24 months	None	0.400 ± 0.07
18—24 months	Irradiated	0.290 ± 0.04
8—10 weeks	None	0.270 ± 0.04
8—10 weeks	Irradiated	0.280 ± 0.04

[a] C57BL/6 mice of the age indicated were untreated or were exposed to 800 rads gamma irradiation with their bone marrow shielded. Six weeks later the mice were immunized with 10 μg TNP-F intravenously. Six days later the mice were bled and their serum assayed for auto-anti-Id by ELISA assay.[7]

Adapted from Kim, Y. T., Goidl, E. A., Samarut, C., Weksler, M. E., Thorbecke, G. J., and Siskind, G. W., *J. Exp. Med.*, 161, 1237, 1985.

Table 9
INFLUENCE OF SPLENIC T CELLS FROM DONORS OF DIFFERENT AGES ON AUTO-ANTI-ID ANTIBODY RESPONSES

Age of irradiated mice	Age of T-cell donor	Average % (± SE) hapten-augmentable PFC
8 weeks	None	17.0 ± 4
8 weeks	8 weeks	16.0 ± 3
8 weeks	24 months	31.0 ± 4
18 months	None	15.0 ± 3
18 months	8 weeks	10.0 ± 4
18 months	18 months	36.0 ± 6

Note: C57BL/6 mice of the age indicated were untreated or were exposed to 800 rads gamma irradiation with their bone marrow shielded. Seven days later they were injected intravenously with 2 × 10⁷ splenic T cells from donors of the indicated age. Seven weeks after irradiation mice were immunized with 10 μg TNP-F intravenously and their responses analyzed 6 days later.

Adapted from Kim, Y. T., Goidl, E. A., Samarut, C., Weksler, M. E., Thorbecke, G. J., and Siskind, G. W., *J. Exp. Med.*, 161, 1237, 1985.

Table 10
ANTISELF-REACTIVITY ARISING IN THE
RESPONSE TO TNP-F IN YOUNG AND AGED MICE

Age	Anti-TNP PFC/ spleen ($\bar{x} \pm$ SE)	Percentage of anti-TNP response reactive with	
		Br-MRBC	M transferrin
6—8 weeks	9500 ± 600	46	31
24 months	4000 ± 300	75	58

Note: C57BL/6 mice of the indicated age were immunized with 10 μg TNP-F administered intravenously. Seven days later single-cell suspensions obtained from their spleen were tested for plaque formation against TNP-SRBC, Br-MRBC, and m-transferrin-SRBC.

Data from Goidl, E. A. and Martin McEvoy, S., *Fed. Proc. Fed. Am. Soc. Exp. Biol.*, 46, 1383, 1987.

old mice may be due to Id-anti-Id interactions which occur between cells arising from the bone marrow and the long-lived peripheral T cells.

V. AUTO-ANTI-ID ANTIBODY AND ANTI-SELF-REACTIVITY IN THE AGED IMMUNE RESPONSE

Within the premise that shifts in clonal distribution of long-lived peripheral T cells are the consequence of life-long interactions with internal and environmental antigens, it is appropriate to ask whether there is a beneficial component to the increased production of auto-anti-Id in the aged immune response. I shall discuss in this section the preliminary data obtained recently in my laboratory. The extent of bystanding activation or cross reactivity towards self-antigens which are generated during the normal immune response of aged mice has been determined. Aged mice have been immunized with 10 μg TNP-F and their immune responses measured at different times following antigen administration. Their anti-TNP PFC responses were measured as well as their capacity for plaque formation against self-antigens. Three such self-antigens were chosen, bromalin-treated mouse erythrocytes (Br-MRBC), mouse transferrin, and mouse albumin. A considerable degree of antiself-reactivity was generated in the immune response of both young and old mice (Table 10). These antiself-reactivities increased dramatically with age: at 24 months of age, 75% of the magnitude of the anti-TNP PFC response was found to cross react with Br-MRBC; similarly, 58% was cross reactive to mouse transferrin. In these series of experiments, the levels of cross reactivity to mouse albumin was always very low (<10%). One should emphasize that at this time it is not possible to distinguish actual cross reactivity of the anti-TNP PFC response from the activation during immunization of bystanding lymphocyte populations with anti-self-reactivity. Such differentiation will be determined at the level of monoclonal antibodies obtained from such immune cell populations.

One finding which supports the proposition that at least a proportion of the anti-TNP response is cross reactive towards self-antigens was obtained following the removal of auto-anti-Id from spleen cells obtained from old mice (Table 11). Another possibility is that the auto-anti-Id produced in the aging response is degenerate.

After removal of auto-anti-Id, antiself-reactivity in the immune cell population obtained from young mice was found to increase modestly (<15%). In contrast, removal of auto-anti-Id from aged immune spleen cells led to an increase in anti-self-reactivity ranging from 28 to 42%. Still, these results are as yet inconclusive in determining the exact origin of the anti-self-reactivities.

Table 11
EFFECT OF REMOVAL OF AUTO-ANTI-ID
ANTIBODY FROM IMMUNE CELL POPULATIONS
FROM YOUNG AND AGED MICE ON ANTISELF-
REACTIVITY

Age (months)	Percentage increase in antiself reactivity after anti-Id removal (individual mice)
2—3	0, 6, 10, 8, 10, 14
18—20	28, 33, 39, 34, 42, 31

Note: C57BL/6 mice of the age indicated were immunized with 10 μg of TNP-F administered intravenously. Seven days later single-cell suspensions obtained from spleen were plaqued against TNP-SRBC and an aliquot was incubated in vitro with 10^{-9} M TNP-epsilon amino-caproic acid to remove auto-anti-Id, and washed before plaquing vs. TNP and vs. self-antigen targets.

Data from Goidl, E. A. and Martin McEvoy, S., *Fed. Proc. Fed. Am. Soc. Exp. Biol.*, 46, 1383, 1987.

VI. CONCLUSION

One of the dramatic changes which occurs in the aging immune response is the increase in auto-anti-Id production. This increased production is under the control of the peripheral splenic T cell: presumably, a long-lived cell which has been influenced by lifelong interactions with internal and external environmental antigens. The capacity to produce high auto-anti-Id once acquired in the peripheral lymphoid system then becomes relatively thymic independent. The role of the increased auto-anti-Id response in aging may be to down regulate anti-self-responses which arise during the immune response. The physiological consequences of such increased auto-anti-Id production may lead to a decrease in the resistance to infectious agents in aging. This is particularly noteworthy, since auto-anti-Id may down regulate high affinity Id which would be more effectively protective.

REFERENCES

1. **Jerne, N. K.,** Towards a network theory of the immune system, *Ann. Immunol. (Paris),* 125C, 373, 1974.
2. **Wigzell, H.,** Positive autoimmunity, in *Autoimmunity. Genetic, Immunologic, Virologic, and Clinical Aspects,* Talal, N., Ed., Academic Press, New York, 1977, 693.
3. **Goidl, E. A., Innes, J. B., and Weksler, M. E.,** Immunological studies on aging. II. Loss of IgG and high-avidity plaque-forming cells and increased suppressor cell activity in aging mice, *J. Exp. Med.,* 144, 1037, 1976.
4. **Jerne, N. K. and Nordin, A. A.,** Plaque formation in agar by single antibody-producing cells, *Science, (Washington, D.C.),* 140, 405, 1963.
5. **Schrater, A. F., Goidl, E. A., Thorbecke, G. J., and Siskind, G. W.,** Production of auto-anti-idiotypic antibody during the normal immune response to TNP-Ficoll. I. Occurrence in AKR/J and BALB/c mice of hapten-augmentable anti-TNP plaque-forming cells and their accelerated appearance in recipients of immune spleen cells, *J. Exp. Med.,* 150, 138, 1979.
6. **Goidl, E, A., Schrater, A. F., Siskind, G. W., and Thorbecke, G. J.,** Production of auto-anti-idiotypic antibody during the normal immune response to TNP-Ficoll. II. Hapten-reversible inhibition of anti-TNP plaque-forming cells by immune serum as an assay for auto-anti-idiotypic antibody, *J. Exp. Med.,* 150, 154, 1979.

7. **Gibbons, J. J., Goidl, E. A., Shepherd, G. M., Thorbecke, G. J., and Siskind, G. W.,** Production of auto-anti-idiotypic antibody during the normal immune response. XII. An enzyme-linked immunosorbent assay for auto-anti-idiotypic antibody, *J. Immunol. Methods,* 79, 231, 1985.

8. **Goidl, E. A., Thorbecke, G. J., Weksler, M. E., and Siskind, G. W.,** Production of auto-anti-idiotypic antibody during the normal immune response: changes in the auto-anti-idiotypic antibody response and the idiotype repertoire associated with aging, *Proc. Natl. Acad. Sci. U.S.A.,* 77, 6788, 1980.

9. **Szewczuk, M. R. and Campbell, R. J.,** Loss of immune competence with age may be due to auto-anti-idiotypic antibody, *Nature (London),* 286, 164, 1980.

10. **Goidl, E. A., Choy, J. W., Gibbons, J. J., Weksler, M. E., Thorbecke, G. J., and Siskind, G. W.,** Production of auto-anti-idiotypic antibody during the normal immune response. VII. Analysis of the cellular basis for the increased auto-anti-idiotypic antibody production by aged mice, *J. Exp. Med.,* 157, 1635, 1983.

11. **Schrater, A. F., Goidl, E. A., Thorbecke, G. J., and Siskind, G. W.,** Production of auto-anti-idiotypic antibody during the normal immune response to TNP-Ficoll. III. Absence in nu/nu mice: evidence for T cell dependence of the anti-idiotypic antibody response, *J. Exp. Med.,* 150, 808, 1979.

12. **Ogden, D. A. and Micklem, H. S.,** The fate of serially transplanted bone marrow cell populations from young and old donors, *Transplantation (Baltimore),* 22, 287, 1976.

13. **Harrison, D. E., Castle, C. M., Doubleday, J. W.,** Stem cell lines from old immunodeficient donors give normal responses in young recipients, *J. Immunol.,* 118, 1223, 1977.

14. **Zharhary, D. and Klinman, N. R.,** B cell repertoire diversity to PR-8 influenza virus does not decrease with age, *J. Immunol.,* 133, 2285, 1984.

15. **Kim, Y. T., Goidl, E. A., Samarut, C., Weksler, M. E., Thorbecke, G. J., and Siskind, G. W.,** Bone marrow function. I. Peripheral T cells are responsible for the increased auto-anti-idiotypic response of older mice, *J. Exp. Med.,* 161, 1237, 1985.

16. **Goidl, E. A. and Martin McEvoy, S.,** Mechanisms of anti-self reactivity in the aging immune response, *Fed. Proc. Fed. Am. Soc. Exp. Biol.,* 46, 1383, 1987.

INDEX

Printed and bound by CPI Group (UK) Ltd, Croydon, CR0 4YY

22/10/2024

01777633-0006